Trusting Science

When to trust scientific claims

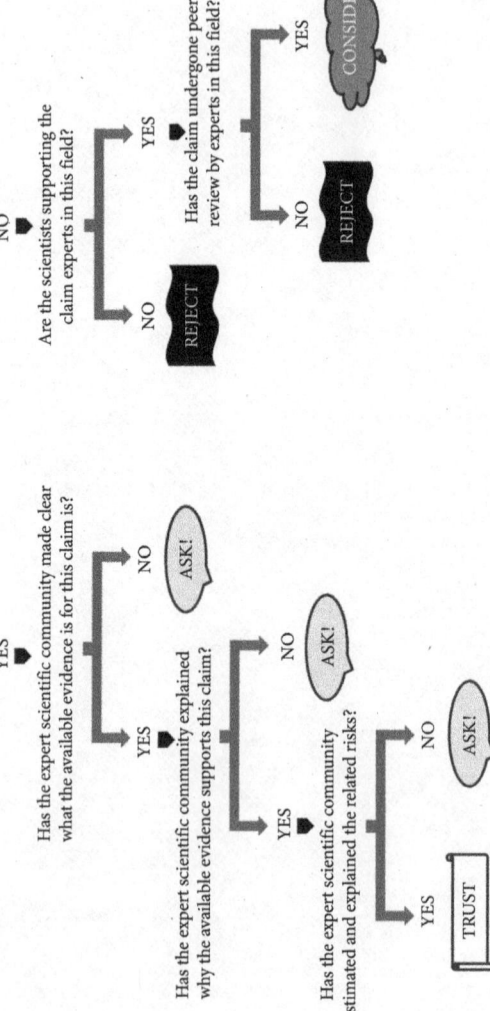

Is the claim related to a scientific question?

NO — ethical question / political question

These are questions informed by science, but ones that it cannot itself answer.

YES — Is there a conspiracy theory proposed?

YES — conspiracy impossible → REJECT

YES — conspiracy possible

NO — Is the claim in agreement with the consensus view of the expert scientific community?

NO — Are the scientists supporting the claim experts in this field?

NO → REJECT

YES — Has the claim undergone peer review by experts in this field?

NO → REJECT

YES → CONSIDER

YES — Has the expert scientific community made clear what the available evidence is for this claim?

NO → ASK!

YES — Has the expert scientific community explained why the available evidence supports this claim?

NO → ASK!

YES — Has the expert scientific community estimated and explained the related risks?

NO → ASK!

YES → TRUST

Endorsements

"In a climate where distrust and misunderstanding of, and misinformation about, science threatens our ability to respond effectively to urgent social, technological and medical crises, this book is sorely needed. *Trusting Science* is a clear and balanced guide that should help anyone make informed decisions about new and perhaps controversial or contested scientific claims, as well as guiding policymakers toward ways of maintaining public trust in science—which, for all its flaws and uncertainties, is still the best system we have for developing reliable knowledge about the natural world."

Philip Ball, science writer and author of How Life Works

"*Trusting Science* is designed to help readers make both personal and societal decisions that are based on the best science. How to sort through all the confusing claims that only pretend to be based on science, including the many 'conspiracy theories' that permeate the internet? Using a series of historical examples, Kampourakis carefully explains the difference between what science can and cannot do, and how societal decisions often depend on questions that science cannot answer. Throughout, he also stresses that there are many important concepts related to how science is done that are never taught in school but must be taught if adults are to decide what science to trust. The entire book can be viewed as preparation for its final, and most important figure: an invaluable personal 'decision tree' about when to trust scientific claims."

Bruce Alberts, Chancellor's Leadership Chair in Biochemistry and Biophysics for Science and Education, University of California, San Francisco, Former Editor-in-Chief, Science magazine (2008-2013), President Emeritus, US National Academy of Sciences (1993–2005)

"In *Trusting Science*, Kostas Kampourakis examines the roots of science skepticism and proposes educational solutions. He usefully draws from historical and contemporary examples, particularly in vaccination, to offer valuable insights into scientific processes, limitations, and the importance of trusting scientific consensus. Kampourakis makes a much-needed call to reform science education to focus on the philosophy and sociology of science rather than on the mere memorization of scientific facts. The book's focus on practical applications will make it valuable for every citizen as well as for educators, policymakers, and everyone who is interested in the important role science plays in society."

Mark Navin, Professor and Chair of Philosophy, Oakland University and author of America's New Vaccine Wars

"Kostas Kampourakis is one of the most important scholars in the world exploring and critiquing the ways that various publics encounter science. His excellent *Trusting Science* is another triumph. Kampourakis offers here a compelling and insightful treatment that all teachers, as well as all citizens who care about getting the teaching of science right, should read."

Robert D. Johnston, Professor of History, University of Illinois Chicago

"In today's world the rising tide of science misinformation and mistrust threatens to drown us all—with disastrous consequences for society. In this brilliant new book, Kampourakis offers a fresh way to think about what science is and why it can be relied on to make decisions about our lives. Essential reading for science educators, policymakers, and anyone who cares about safeguarding our future."

John L. Rudolph, Vilas Distinguished Professor, University of Wisconsin-Madison and author of Why We Teach Science (and Why We Should)

"A healthy dose of skepticism can be a good thing. But what are we to do when the dosage of skepticism threatens an overdose? This is the situation we find ourselves in when it comes to many people's views of science. There is a deep distrust of the findings of science and the reports of scientists— especially when it comes to things like climate change and vaccines. *Trusting Science* is a much-needed correction to the overdosing of science skepticism prevalent today. In this book Kostas Kampourakis skillfully explores the underlying causes of much skepticism about science. He not only tracks the historical development of science skepticism of various sorts, but he also examines the causes with an even hand noting where some skepticism is indeed warranted and where it is unfounded. *Trusting Science* offers readers a chance to better understand why there is so much distrust of science and what might be done to dispel it. This book is clear, persuasive, and important. A must read!"

Kevin McCain, Professor and Chair, Department of Philosophy, The University of Alabama at Birmingham

Trusting Science

Why We Need to Reconsider School Science Teaching

Kostas Kampourakis
University of Geneva

OXFORD
UNIVERSITY PRESS

OXFORD
UNIVERSITY PRESS

Oxford University Press is a department of the University of Oxford.
It furthers the University's objective of excellence in research, scholarship,
and education by publishing worldwide. Oxford is a registered trade mark of
Oxford University Press in the UK and in certain other countries.

Published in the United States of America by Oxford University Press
198 Madison Avenue, New York, NY 10016, United States of America.

Library of Congress Cataloging-in-Publication Data
Names: Kampourakis, Kostas, author.
Title: Trusting science : why we need to reconsider school science
teaching / Kostas Kampourakis.
Description: New York, NY : Oxford University Press, [2025] |
Includes bibliographical references and index.
Identifiers: LCCN 2024062157 (print) | LCCN 2024062158 (ebook) |
ISBN 9780197787106 (hardback) | ISBN 9780197787120 (epub) | ISBN 9780197787137
Subjects: LCSH: Science–Social aspects. | Science–Study and teaching. |
Trust–Social aspects. | Denialism. | Expertise.
Classification: LCC Q175.5 .K28 2025 (print) | LCC Q175.5 (ebook) |
DDC 507.1–dc23/eng/20250310
LC record available at https://lccn.loc.gov/2024062157
LC ebook record available at https://lccn.loc.gov/2024062158

DOI: 10.1093/oso/9780197787106.001.0001

Printer by Marquis, Canada

To all those teachers who struggle every day to provide their students with a deeper understanding of science, not just content knowledge or rote learning

Contents

Preface

Should we trust science, and to what extent? Or should we question its findings? Is skepticism towards science a healthy attitude, or should we unquestionably accept its conclusions? Perhaps you have already wondered whether science is as trustworthy as we often take it to be. Or perhaps you have seen other people question its validity and reliability. Perhaps you are thinking that you already have the foundations to understand how science is done and decide for yourself how trustworthy it can be. Or you may think that you do not know much about science and thus decide to trust it blindly in the same way that you trust the pilots who fly the airplanes in which you travel.

If you did not have any formal science education after leaving school, and if you are trying to decide when to trust science, then this book is for you.

For the purposes of the present book, science is defined as the systematic empirical study of the natural world with the aim to understand it. There are many important concepts related to how science is done that you were never taught in school. These are as important as scientific knowledge itself, and they become even more important when it comes to deciding whether or not to trust science. The present book fills this gap and offers lay readers a guide to a better understanding of what science is, how it is done, and what its limitations are. It is only then, I argue, that you can confidently decide when science can be trusted. The heuristic tool at the beginning and at the end of the present book, which combines the various aspects of science discussed in it, can be a useful guide for deciding when to trust scientific claims.

For the past 10 years, my colleague Bruno J. Strasser and I have been teaching a course called "Biology & Society" to second-year undergraduate students in the life sciences. By the time the course begins, these students have a solid knowledge of biology and its scientific practices, previously acquired during an intensive first year of their undergraduate studies. In general, most of them have appreciated the socio-scientific controversies that we have taught in our course. But I have come to realize that they also have many questions beyond the topics we have been discussing, which often are questions about the nature of science itself. Some of them were answered during the course, while others were not. This made me think that if these biology undergraduates are left with questions about the nature of science unanswered, what about all those others, the lay people out there, who have not had any formal education in science after school? These are often referred to as "the public," and should be able to participate in decisions, directly or indirectly, as citizens in democratic societies. It is all these lay people that I had in mind while writing the present book.

But what we mean by "public" is neither straightforward nor self-evident. In the present book, I use this word as a noun or an adjective vaguely to refer to all lay people, that is, all people who are not experts in science. I must note though that there is no single group of nonexperts that we can define as "the" public, as people around the world differ in their perceptions, understanding, and knowledge of science, depending on their cultural contexts. Therefore, it is better to refer to "publics." But to make sense of who these "publics" are requires a clarification of who an expert is in science. Both "experts" and "publics" are complex categories that depend on the context. These categories have also changed over time, depending, on the one hand, on the level of experts' knowledge and understanding of the natural world and, on the other hand, on publics' attitudes towards that knowledge and understanding. However, we can roughly make a distinction between experts and nonexperts in a particular scientific field: Experts are those people who have specialized knowledge and skills related to that field, who practice these as their main occupation, and who have valid credentials and affirmation by their peers.

This contrast between experts having specialized knowledge and skills that the publics lack has long supported the "deficit model" of the relation between experts and publics. According to this model, scientific knowledge and understanding are transmitted by the knowledgeable experts to the ignorant publics, in an attempt by the former to educate the latter. In this view, experts always have superior status compared to publics. Yet, this is far from accurate. Both the way science itself is conducted and the way its findings are communicated have never been completely separated from their social contexts. Therefore, the communication of scientists' conclusions to the various publics is not a linear process of transmission, but one of constant interaction and negotiation. This entails that by stating my intention to offer readers an opportunity for reflection about science that they may have not had in school, I am not relying on any deficit model. What I am offering is not the enlightenment of the ignorant but an opportunity for reflection they may have never been given before.

When the recent COVID-19 pandemic came, questions about trust in science came to the fore. The public discussions during the pandemic as well as my own foray into the history of vaccination for our "Biology & Society" course made me discern particular underexplored causes of public distrust in science that all have to do in one way or another with what science is, how it is done, and what its limitations are. These causes are the focus of the present book, which aims to offer lay readers an opportunity for reflection about science that they may have never had in school, in order to help them understand what science is and when it should be trusted. Each chapter considers particular episodes and draws lessons from them to provide readers with opportunities for reflection and some useful thinking tools about the respective issue. A secondary aim of the present book is to showcase the need for a reconceptualization of science teaching in schools. This is why the concluding section of each chapter in the book is titled "What is crucial to understand? (and also teach in schools!)." Although I am primarily interested in promoting reflection among the readers of the book, there is a lot of food for thought for policymakers and curriculum developers.

Even though it is evident in the table of contents that vaccination has a central place in the present book, this is not another book about vaccination or the vaccination controversies per se. It does not provide a global or local history of vaccination and of the reactions towards it, and it is not a book in academic history of science, neither in terms of methods nor in terms of sources. Neither does the book aim to discuss in detail the various medical, sociological, psychological, anthropological, financial, and other issues that pertain to vaccination. Rather, this is a book that just draws on the history of vaccination controversies to identify previously overlooked reasons for why people may distrust vaccination and the science behind it. This is why in each of the chapters of the book other topics are also considered, such as syphilis (Chapter 3), embryo ethics (Chapter 4), cancer (Chapter 5), AIDS (Chapter 6), scientists' unconscious biases (Chapter 7), religion (Chapter 8), eugenics (Chapter 9), and evolution (Chapter 10). The history of vaccination is just the point of departure. The causes for distrust thus identified in each chapter also exist elsewhere in science, as the cases discussed in each chapter clearly show.

But to reveal these causes of distrust, my coverage of history had to be selective rather than exhaustive. One limitation of the present book is therefore its coverage, as the stories presented come from two countries mostly: the USA and the UK. This choice was made for pragmatic reasons. One is the availability of historical analyses and evidence in these two countries. Another is the direct accessibility of all the relevant material in English. I strongly believe that all lessons drawn from the various episodes considered herein are generally applicable, at least in the western world, which is the social and cultural context I live in and am familiar with.

I hope that you will find the present book interesting and useful.

Acknowledgments

I am indebted to Jeremy Lewis, my editor at Oxford University Press, who supported the idea for the present book right from the start and until its publication, as well as to his colleagues who worked with me towards the publication of the present book: Madeline Hoverkamp, Jacqueline Ridberg Larabee, Hemalatha Ravivarman, and Praveen Corera.

I am also indebted to my colleague Bruno J. Strasser for his support and the opportunity to teach with him the course "Biology & Society," which was his idea in the first place. It was my classes in this course that made me closely study the history of vaccination and "discover" the causes of science distrust considered herein.

Several people provided very helpful comments and suggestions while I was writing this book, and I am very grateful to them all: Philip Ball, Robert D. Johnston, Alan C. Love, Kevin McCain, Mark Navin, Angela Potochnik, and Michael Reiss.

I also owe gratitude to those who provided me with valuable materials for the book: Natalie Snoyman, Supervising Archivist, Lucretia Little History Room, and Mill Valley Public Library for the photograph in Figure 2.1; Julie Anne Lambert, Librarian of the John Johnson Collection for her help in acquiring the photograph in Figure 4.3. It is also amazing how much useful historical material, such as images and books, one can find on the webpage of the Wellcome Collection.

I was a schoolteacher for 12 years and I always tried to offer my students more than content knowledge and rote learning. Now that I am a teacher educator, I try to help teachers do the same. It is not easy to do, and there exist many teachers who struggle every day to offer their students a deeper understanding of science. The present book is dedicated to them, in the hope that it might also become a useful tool in their work, even though it is primarily intended for interested lay readers.

1

In science we trust?

What science is, and is not

What is science? You were likely taught science classes in school, and you may remember those either with amazement or with horror. We all believe we know what science is, but if you think about it, it is far from simple and straightforward to define "science," or to concisely articulate what it is about. The term is generally used to refer to a variety of theories, models, explanations, concepts, practices, attitudes, approaches, and more, that relate to the study of the natural world. Philosophers have shown that there is no single entity to refer to as *science*; instead, there are a variety of disciplines that study the natural world with a variety of methods and explanatory aims. Therefore, it is more appropriate to refer to "sciences," in plural, in order to account for this variety. There are other kinds of science too, and a distinction is often made between the natural sciences and the social sciences. However, in the public sphere, the singular term is commonly taken to only refer to the sciences that study the natural world. As we see in this chapter, various polls try to measure attitudes towards science and scientists without even specifying what these terms refer to, seemingly taking for granted that reference is made to the study of the natural world.

Elsewhere, based on a detailed consideration of a particular scientific theory, I have proposed the following eight basic features of science as a means to studying and understanding the natural world:

1. All scientific conclusions rely on inferences from empirical data.
2. These conclusions must be confirmed by independent testing.
3. If several conclusions are supported by empirical data and by independent testing, then scientists can accept the one that provides the best understanding.
4. It is the collective understanding of the natural world, and not truth in any absolute sense, that is the ultimate aim of science.

Trusting Science. Kostas Kampourakis, Oxford University Press. © Oxford University Press (2025).
DOI: 10.1093/oso/9780197787106.003.0001

5. The collective understanding that science produces is summarized in the form of scientific theories that rely on abstractions and idealizations (models) of the natural world.

6. Scientific theories are open to revision or rejection in the light of new evidence and new inferences that might bring about new understanding.

7. Because of its reliance on empirical evidence and independent testing, and its openness to revision, science provides the most objective rational means of understanding the natural world.[1]

It will be useful to keep these features in mind while reading the present book, as a reminder of what science can do and how it does it. The list is not exhaustive, but I believe that it contains some of the most important features of science as the enterprise that aims to understand the natural world through its systematic empirical study.

Among these important features, it is worth emphasizing the first one because it forms the basis for all the others that follow: that science is based on *inferences* from *empirical* data. These data in turn can become evidence that can be used to support or reject a particular hypothesis, or to simply answer a research question in cases where a hypothesis has not been formulated in advance—even though, in general, scientists formulate hypotheses and put them to the test (see Box 1.1). I have written that "data" can be used as "evidence" because these terms are not synonymous. Data are just facts; they have no meaning on their own. When data are studied and interpreted, they can become evidence, that is, a reason to arrive at a particular conclusion, or to use them in favor of or against the justification of a belief. This entails that the same data can be used as evidence for arriving at different conclusions or supporting different beliefs, and therefore that it is not the data themselves but their interpretation that matters. In short, contrary to claims made in TV series and elsewhere, the data do not speak for themselves; who interprets them and how can make a huge difference.

Box 1.1 Hypothesis testing and inferences from data

In science, it is possible to propose hypotheses and test them by collecting data from experiments, observations, or other procedures. There are at least two alternative hypotheses for a given situation: either that a relation does not exist (this is often described as the "null" hypothesis) or that a relation does in fact exist. This can be made clearer by using a legal analogy:[2]

- The suspect is innocent, or there is no relation to the crime (null hypothesis).
- The suspect is guilty, or there is a relation to the crime (alternative hypothesis).

In this case, there are two kinds of errors that are possible:

- *Type I error*: Reject the null hypothesis when it is true, that is, find a relation when it does not really exist; this entails wrongly convicting an innocent person.
- *Type II error*: Do not reject the null hypothesis even though it is false, that is, do not find a relation although it does exist; this entails wrongly failing to convict a guilty person.

These different outcomes can be summarized in the following table:

	Do not reject the null hypothesis (find suspect not guilty)	Reject the null hypothesis over an alternative (find suspect guilty)
Null hypothesis (suspect is innocent)	Correctly not reject the null hypothesis (find an innocent person not guilty)	Incorrectly reject the null hypothesis, or make a Type I error (convict an innocent person)
Alternative hypothesis (suspect is guilty)	Incorrectly fail to reject the null hypothesis, or make a Type II error (fail to convict a guilty person)	Correctly reject the null hypothesis (convict a guilty person)

It is in a similar sense that in science we can make inferences from data about whether or not particular relations exist. For instance, when epidemiological studies investigate relations between a factor and a disease, there are four possible outcomes:

- A true relation is inferred (true positive).
- A relation that does not really exist is inferred (false positive).
- A true relation is not inferred (false negative).
- A relation that does not really exist is not inferred (true negative).[3]

		Relation	
Statistical		True (Exists)	False (Does not exist)
inference	Yes (Inferred)	True positive	False positive
from data	No (Not inferred)	False negative	True negative

Therefore, based on the available data, we can correctly find that a relation does and does not exist, but we can also be wrong. This awareness that we can make wrong inferences is the basis of the self-correcting nature of science.

By accumulating data, analyzing, and interpreting them, scientists study the natural world and try to understand it. There are two types of understanding in science: understanding a theory, that is, being able to use it; and understanding a phenomenon, that is, having an adequate explanation of it. Scientific understanding thus arises when scientists understand a theory and then use that understanding to generate adequate explanations and predictions of phenomena. For a scientific theory to provide genuine understanding in science, it must agree with the empirical data collected by observation or measurement, not contain contradictory elements, and be used effectively in order for scientists to be able to make sense of the natural phenomena they study.[4] It is important to note here that the word "theory" has an entirely different meaning in science compared to its colloquial use, where it is used to denote a hypothesis, a thought, or a speculation. A scientific theory is a collection of models, principles, and assumptions that can provide explanations and predictions about natural phenomena. From a rational and naturalistic point of view, our understanding cannot get any better than what our current theories allow for, and this is why there exist limitations to our understanding.

Let us consider by way of example the scientific theory that is relevant to many of the topics discussed in the present book. This is the germ theory of disease that explains how particular infectious diseases are primarily caused by the transmission of particular microorganisms (which are described as pathogens because they cause disease) from one host to another. This theory—in fact a collection of such theories—was proposed during the late nineteenth century by laboratory scientists such as Louis Pasteur and Robert Koch, with crucial contributions from pathologists, botanists, experimental physiologists, chemists, and veterinary scientists. Initially, there were concerns about the theory, but Koch's research on the anthrax bacterium in the late 1870s provided clear evidence about the relation between microorganisms and disease. Koch managed to show that the bacterium he had considered as the causal agent existed only in animals with the disease, and it could produce the disease in healthy animals when it was transferred to them. The importance of the germ theory consisted in bringing about a reconceptualization of nature that both changed how people behaved in their everyday lives and allowed for large-scale public health interventions. Until that time, there were various hypotheses about the origins of diseases; it was thought that they were caused by rotting human, animal, or plant filth that escaped into the air and produced miasmas, that is, emanations that caused disease (which is why this is often described as the miasma theory of disease). But once it was accepted that microorganisms were the cause, measures, such as sanitation and vaccination, could be taken to block their transmission.[5]

An important feature of scientific theories is that they evolve as new data become evidence for differing views. During the mid-twentieth century, it became evident that whereas the same pathogen was supposed to cause the same disease, not all people infected by the same pathogen experienced the disease in the same way. Other factors such as their genetic constitution, everyday habits, and socioeconomic status also seemed to play a role, making the notion that infectious disease is primarily caused by a pathogen transmitted from one host to another an oversimplification. This did not mean that the germ theory of disease was wrong, as disease was not possible in the absence of infection by the respective pathogens. But at the same time, the germ theory could not explain all the observed variation in how disease developed.[6] More recently, it has become clear that this also depends on a person's immune system. Therefore, even though the pathogens remain an important cause, they are not the only cause of the disease. It is indeed the case that if there is no pathogen, there will be no disease. However, for those patients who die from infection, the cause of death eventually also depends on their own physiology. A complete theory should therefore account for the observed variation in disease severity among individuals, including asymptomatic carriers, which indicate the important role of host variability.[7] As is well known, many people were infected by SARS-CoV-2, but not all of them experienced the disease in the same way. That some died whereas others did not even realize they were infected was mostly due to their own immune system, physiology, and health condition.

This brings us to a final important point. Scientists do not always have answers to all questions. During the COVID-19 pandemic, it was not possible to explain why some people died whereas others did not. This is why, alongside the key features of science already considered, it is equally important to keep in mind its limitations. Some proponents of science have argued, however, that science is the only way of knowing in general, an attitude described as scientism. But science is a human activity and therefore it is limited by human perception. As we may not be able to perceive everything that may exist, we may not be able to put it to the test. Therefore, science may not be able to provide answers to all questions we would like to answer. Because of this, we should not expect from science more than what it can actually do.

Philosopher of science Susan Haack has noted that there is no simple way to determine whether the line between an appropriate respect for the achievements of science and an inappropriate deference to it has been crossed. However, she has provided a list of what she considers as characteristic indicators of this crossing, each of which denotes some misunderstanding of science:

1. Forgetting that science makes claims, many of which will not survive the test of time, and being ready to accept anything that bears the label "scientific."
2. Using the terms "science" and "scientific" as synonyms for "strong, reliable, good."
3. Insisting on a sharp line of demarcation between science and nonscience.
4. Considering that what is distinctive about science is its method, which is supposed to be unique and distinct from other forms of inquiry.
5. Adopting the tools of science to disguise a lack of real rigor.
6. Attempting to take over nonscientific disciplines and replace them with science.
7. Denigrating the importance, or even denying the legitimacy, of nonscientific disciplines.[8]

This is another list that is useful to keep in mind while reading the present book, as a reminder of what science can do and how it does it. Scientism raises high expectations from science; when science fails to fulfill those high expectations, we may perceive it as inadequate. In a book about science, it is therefore necessary to be clear about what science can and cannot do.

To summarize: Science is a means to understanding the natural world; its theories can and should be put to the test and be confirmed independently or revised; above all, science is not something to which we should dogmatically subscribe. Because of all these characteristics, science provides the best rational understanding of the natural world possible. But then why do some people question, or even deny, the claims of science?

Denying science

It is common nowadays to describe resistance to the claims of scientists as science denial, which can be defined as a motivated rejection of science when it does not support one's prior beliefs. In general, being skeptical about the claims of scientists and trying to examine the relevant evidence is legitimate and, in fact, an attitude inherent in the processes of science. No serious scientist would ever complain if they were asked to show the evidence on which their claims are based. However, this is not what science deniers do. Rather, they outright reject the claims of scientists because they prefer to maintain their own beliefs rather than seek and evaluate the scientific evidence for a particular claim.[9] Science denial is an attitude often related to science doubt, which is the motivated and conscious effort to question the claims of scientists in order to serve vested interests. One well-documented case is how the tobacco industry in the 1970s tried to emphasize the uncertainties around the connection between tobacco smoking and lung cancer, in order to raise doubt about the conclusions of scientists regarding the existence of this connection that had been observed already in the 1930s in Germany.[10] The common aspect in both science denial and science doubt arguments is that mainstream scientists are wrong, and so their claims are to be rejected. Science denial is an explicit rejection of science; science doubt is not as explicit, but its motivation and ultimate goal is the rejection of science too. For this reason, hereafter I refer to both the explicit (denial) and implicit (doubt) rejection of science collectively as science denial.

For the purposes of the present book, we can define science denial as the systematic rejection of scientists' conclusions that is not based on a careful and objective consideration of the available data. This is an attitude whereby people consciously overlook the suggestions of the scientific community and oftentimes believe that they themselves hold the truth that scientists either miss or want to hide. Science denial is in fact a political stance that explicitly challenges the claims made by the scientific community. But this is not all. In a thoughtful analysis, philosopher Lee McIntyre has suggested five specific features of science denial: cherry picking of evidence, belief in conspiracy theories, reliance on fake experts, committing logical errors, and setting impossible expectations for science. These operate in concert with one another when science deniers use them to make their case that whatever scientists claim is incorrect. But McIntyre also emphasizes that this is just the argumentative strategy of deniers; it does not explain why they are, or how they become, deniers in the first place. Rather, the reasons that they deny science are psychological and have to do with who they are. This is why providing them with more and more scientific evidence is unlikely to change their minds; for this to happen, they would have to undergo a change in their social identity. McIntyre suggests that if we intend to convince them to better consider the views of mainstream science, we need to approach them with respect, earn their trust, and ask them questions that will really make them reconsider things, such as what kind of evidence would make them change their minds. But even this might not work.[11]

A qualification is necessary here. When we refer to science denial in general, one might get the impression that all science is denied. However, this is very far from true. In contrast, there exist whole scientific fields that are not disputed, questioned, or denied by lay people. For instance, not many people, if any, question quantum mechanics, solid-state physics, plate tectonics, atomic theory, or the fundamentals of molecular biology. Indeed, if we pay attention to what science denial often focuses on, it is usually what we can describe as "policy science," that is, particular science disciplines and research programs on which we rely to arrive at policy recommendations, such as climate science or vaccine science. Law and policy do not rely on every scientific discipline, but only on those that are more relevant than others for making decisions and proposing regulations; these are the ones that we can think of as "policy science": science on which policy relies. If you think about it, the denial, or dispute, of particular science disciplines is not independent of policy decisions; this is why it is usually policy science that is contested and denied by lay people.[12] Among the various kinds of policy science, it is vaccine science, or vaccinology—a term introduced by Jonas Salk in 1977 to describe "The study and application of the basic requirements for effective immunization ..."[13]—that has a central place in the present book.

This in turn brings us to a dilemma. When "policy science" is predominantly about policy and less about science, shouldn't it be best subjected to a robust democratic debate with many legitimate valid viewpoints, rather than relying on one "correct" scientific viewpoint? Or because the policy is about a science-related area such as vaccination, should we better trust the scientists to decide? Consider some of the questions that lay people often ask about vaccination: Who should be vaccinated? Should vaccination be compulsory? How long should vaccine development take? How can we establish the efficacy and the effectiveness of a vaccine? What percentage of the national budget should go to that area? Not all of these are scientific questions. Furthermore, even for those questions that are scientific, scientists are not always able to provide definitive answers, and might not even agree among themselves depending on their political or social views.[14] I am inclined to think that the answers to these questions might be informed by science but can in no way be answered on scientific grounds alone. The link between science and policy is both sensitive and fragile.

Turning back to science denial, how prevalent is it really? A recent study investigated the prevalence in 24 countries of the "systematic and unwarranted rejection of science," which the researchers called "science skepticism." However, as already mentioned, being skeptical about the findings of science is absolutely legitimate if one is ready to examine the available evidence in an unbiased manner. I think that the "systematic and unwarranted rejection of science" is better described as science denial, and this is what I consider that they actually measured. What the researchers did was to ask participants to give a score between 1 and 7 to statements such as "Human CO_2 emissions cause climate change," "Vaccinations cause autism," "Genetic modification of foods is a safe and reliable technology," and "Human beings, as we know them today, developed from earlier species of animals," with higher scores indicating rejection of the respective scientific view (note that the statement on vaccination does not express the scientific view, whereas the others do). The average scores were 2.45 for climate change denial, 1.86 for vaccine denial, 4.57 for genetic modification (GM) denial, and 2.74 for evolution denial. As is evident, vaccines are the topic for which there was less resistance in relative terms, compared to the other three topics. However, science denial varied with respect to the country and the topic. There was another interesting conclusion drawn by the researchers, that lacking content knowledge about science, such as that "Electrons are smaller than atoms," is not the main driver of science denial (in that study, this was the case only for GM). They noted that this confirmed previous research that merely addressing knowledge deficits was not sufficient to address science denial.[15]

What conclusions can we draw from these findings? The first is that science denial may not be the most important problem today's societies face. It is particularly interesting that the rejection of vaccination was relatively low with a mean of 1.86 in the range between 1 and 7. Therefore, science denial may not be the main reason that the recommendations of scientists are rejected. Rather, I suggest that science denial is a small-scale movement that manages to have a large impact because it can have a negative influence on many other people who are not themselves denialists but who are influenced by the denialists' views, thus resulting in not trusting science. A second conclusion from the study seems to also support the main argument of the present book, that more than science content knowledge is required for a robust understanding of science. Rather, I suggest that we can enhance trust in science by helping lay people gain a better understanding of science.

Before we get into this, it is necessary to clarify what trust in science is about.

Trust, mistrust, and distrust

What is trust? At its fundamental level, it is about reliance. I rely on the computer I am now using to write to automatically save and preserve my work, thus not worrying about losing it. I rely on the desk I have placed my computer on to not collapse while I am working, thus not worrying about my computer getting damaged. But trust is more than this reliance; it is also about both my expectations and my reaction when these are not met. I cannot say that I trust my computer or desk because I cannot blame them if they fail to meet my expectations; I simply rely on them. The ones I trusted when I bought both of these were their manufacturers. Therefore, if my computer or desk fails to do what I expect from them, it is their manufacturers who are not trustworthy. But for what? When we trust people, we rely on them to meet their commitments. Those who do are considered trustworthy, whereas those who do not are considered untrustworthy. This in turn depends on their intentions and skills. A manufacturer is trustworthy when they have the intention to make good quality products and the skills to do so. In general, when we trust someone, we rely on them to have both the intention and the skills required to fulfill their commitment to us.[16]

In a thoughtful analysis, political scientist Patti Tamara Lenard has nicely distinguished between trust and the different varieties of lack thereof.[17] The reliance on someone having the good intention and the skills required to fulfill their commitment to us makes us vulnerable to the possibility of disappointment because we cannot be certain whether they will indeed do so. To give some examples, we trust elected officials to serve the common good, or judges to deliver justice; we trust the mechanic to repair our car, and the postman to deliver our mail and orders. We also trust our fellow citizens to follow the same laws and rules that we follow in everyday life, for instance, respecting the traffic code to avoid accidents. In all these cases, our trust makes us vulnerable because corrupt officials and judges, incompetent mechanics and postmen, or disobedient fellow citizens might exist. The less corruption, incompetence, or disobedience we encounter, the more we trust the respective people and the related institutions. When these people deliver what we expect them to, we consider them worthy of our trust, and we are ready to trust them again in the future. The more reliably a person does what we expect them to do, the more trustworthy we consider them.

Accordingly, the less reliably a person fulfills their commitment, the less we are going to trust them. In such cases, we may find ourselves in one of two situations that are necessary to clearly distinguish from each other. One is distrust, which is about having suspicion or doubt that particular people can be trusted. This is an attitude that often stems from past experiences: The people we distrust have disappointed us in the past, and because of this we no longer consider them trustworthy. It is important to note that distrust is not simply the lack of trust; rather it is the feeling of suspicion or doubt towards some people. Distrust is different from simply not trusting someone just because we are not in a situation where we would have to do so; for instance, one who never uses taxis does not have to trust taxi drivers, but this is not distrust. Rather, we distrust someone because we suspect that they are untrustworthy due to previous experiences. For instance, one may distrust taxi drivers when one has been repeatedly charged by them more than what they should have been. In general, distrust is accompanied by the belief that most interactions with a particular person, or a particular group, are more likely to end in disappointment rather than success.

Given that we may trust one's competence and one's intentions, there can be two different kinds of distrust. On the one hand, we may distrust a person because we do not consider them competent to carry out some task (e.g., repair our car), even though we know that they have the best intentions to do so (e.g., think of an uncle who is not a mechanic but who claims to know a lot about cars and who wants to help you). On the other hand, we may distrust a person because we do not believe they have good intentions towards us, even though they may be competent to do what we expect them to do (for instance, a skilled mechanic who has had a huge fight with that uncle of ours and does not behave in a friendly manner to the whole family). We can also distrust someone based either on knowledge—we distrust someone because we know that they have disappointed us in the past—or suspicion—we distrust someone because we have been told by others of their being untrustworthy and it is therefore better for us to be cautious in dealing with them. In short, distrust is a conscious lack of trust in someone for a reason.

A related, but different, attitude is mistrust, and it differs in at least two important respects from distrust. First, whereas distrust is a stable and long-term attitude, mistrust is a more tentative one and depends on the context. Second, whereas distrust is characterized by suspicion, mistrust is rather characterized by a well-intended skepticism. Therefore, mistrust is likely to result in ambivalence about the trustworthiness of others, rather than in a perception that they cannot be trusted at all. A mistrustful person is characterized by uncertainty and indecision and may fluctuate between trusting and not trusting someone. This is why additional information can make a big difference for mistrust, shifting one's attitude from being cautious to trusting despite having insufficient evidence. In contrast—and this is a key difference—distrust is resistant to new information that might show that such an attitude may not be warranted. When it comes to healthcare, there are many questions to consider. Do lay people trust the institutions that provide it? If they do not, do they have a general sense of unease towards these institutions (mistrust), or a more active stance against them (distrust)? Most importantly, what can these institutions that are perceived to be untrustworthy do about it in order to regain lay people's trust?[18]

Based on these considerations, there is an important difference between science distrust/mistrust and science denial. Science denial is a conscious and consistent attitude that denigrates science and scientists, in which science is questioned a priori: The reasoning is based on theoretical grounds, rather than on empirical observation and experience. Science denialists are convinced that scientists are wrong and look for evidence to support this conviction.[19] In contrast, science distrust/mistrust is the feeling that people cannot depend on scientists with full confidence. For some people, this is an awkward feeling because they end up not trusting science, even if in principle they might be willing to do so. The key difference with science denial is that in science distrust science is questioned a posteriori: The reasoning is based on experience, such as known facts or past events, rather than theoretical assumptions. People who distrust science do so because they have had an experience where they felt that science failed them. They may thus become prone to be influenced by the arguments of science denialists.

Based on what has been discussed so far, we can now summarize the different possible attitudes towards scientists:

- *Denial*: The explicit rejection of scientists' claims and recommendations, by directly denying those. This includes *doubt*, which is the implicit rejection of scientists' claims and recommendations by openly questioning those.
- *Distrust*: The attitude of being suspicious of scientists' claims and recommendations, by questioning their competence or their motivations.
- *Mistrust*: The attitude of being uncertain and undecided about scientists' claims and recommendations, thus looking for additional information in order to decide whether to trust those.
- *Trust*: The attitude of valuing scientists' claims and recommendations because we trust their intentions and competence.

But what exactly do we trust scientists for, and why?

Expertise and trustworthiness

Simply put, we trust scientists to understand the natural world and either explain it to us or advise us on what to do whenever natural phenomena may impact our lives (for instance, a pandemic or a natural disaster). What makes scientists trustworthy is their expertise. In everyday life, we consider an expert any person who knows how to do something well. One can have expertise in gardening, cooking, or driving. But we should distinguish between ubiquitous expertise, that is, the expertise that we have acquired during our life without any self-conscious effort, and specialist expertise. Gardening, cooking, or driving, or even the language we learned to speak as we grew up, are instances of such ubiquitous expertise. Any person who can do any of these well and better than most others can be considered an expert in doing these. But this is different from being a gardening, cooking, or driving professional. When it comes to a person's profession, whatever that is, it is specialist, not ubiquitous, expertise we are talking about. People are specialist experts in the content of their profession, which is something that can only be achieved with special practice.[20]

When it comes to science in particular, we consider as scientific experts those people who have mastered science-related knowledge and skills, who practice these as their main occupation, and who have valid science-related credentials, confirmed experience, and affirmation by their peers. Even though members of the public may occasionally be involved in science-related activities, they do not do this as their main occupation and also lack the knowledge, skills, credentials, experience, or affirmation by their peers that scientists have.[21] Of course, scientists are not infallible, nor is it true that all of them are very good at what they do. However, at any given time when it comes to understanding nature and dealing with natural phenomena affecting our lives, such as a pandemic or climate change, expert scientists are the ones who understand the respective situation better than anyone else and who can reliably tell us what we had better do and not do. This entails that they are less likely to reach wrong conclusions when it comes to their areas of professional expertise than nonexperts.

This brings us to a point of utmost importance. Scientists are experts in very specific areas only. Outside those, they are likely no different than lay people. Thus, a pediatrician is an expert with respect to child diseases and vaccines but is no more expert than you or me when it comes to climate change. Accordingly, a seismologist is an expert with respect to plate tectonics and earthquakes but is no more expert than you or me when it comes to vaccines. More generally, *a scientist is not, and cannot be, an expert in everything about science*. No serious scientist would claim otherwise. But some have done so, and they have thus misled lay people, affecting their trust in science. The classic example has been described in detail by historians of science Naomi Oreskes and Erik Conway. Particular scientists who were well known and highly respected by virtue of their earlier work in the weapon programs of the Cold War misled the public about the connection between tobacco and cancer, ozone depletion, and more. Motivated either by monetary gain, their political leanings, or both, they presented themselves as experts on topics on which they really had no expertise and questioned the conclusions of the real experts. This they did by emphasizing the uncertainties and suggesting that the related science was still unsettled, for instance by questioning the claim that tobacco smoking causes lung cancer because most people who smoke do not develop it.[22] Scientists are experts on particular topics, and it is only with respect to those that we should trust them; if they claim otherwise, we should be wary of them.

In today's technology-dependent societies, we rely on expertise more than any time before in human history. There is so much knowledge available and most of it is so specialized that there exist very few things that we can actually do on our own. For this reason, we have to rely on experts of various kinds for almost everything. When it comes to science, the most widely discussed topic that pertains to our reliance on experts is climate change. The reports on climate change by the Intergovernmental Panel on Climate Change, the international body for assessing the data related to climate change that was set up in 1988 by the World Meteorological Organization and United Nations Environment Programme, have suggested that humans clearly have a negative impact on the climate. Recent anthropogenic emissions of greenhouse gases—such as carbon dioxide (CO_2) and methane (CH_4)—are higher now than ever before. The resulting warming of the climate is unequivocal, and the changes that this has brought about are significant. The atmosphere is getting warmer, and the sea level is rising at an alarmingly fast pace. Two main causes of this problem are the burning of fossil fuels, which is the main source of the CO_2 released into the atmosphere, and cutting down forests in order to use the land for agriculture or for building cities, which has resulted in less areas being covered by plants that absorb CO_2 during photosynthesis.[23] The evidence necessary to arrive at these conclusions can only be gathered and analyzed by expert climate scientists, on whom we all rely for the conclusions. Their expertise is the basis for our trust in them to arrive at legitimate conclusions. But not only for us lay people.

Philosopher John Hardwig has argued that trust is also important within science, and often more fundamental than empirical data or logical arguments, because these are available only through trust. The reason for this is that science is done by research groups rather than individuals. Therefore, scientists not only have to draw on the work of their predecessors, but they also crucially depend on the work of their collaborators. This dependence becomes even more important and more significant when the research requires the contributions of many people working at the same time to be completed within a reasonable time, as well as when the various members of the same research group have different kinds of expertise. In both cases, each member must have trust in what the others do: that they do what they must do because one cannot do the work of two people at the same time, and that they do correctly what they do because one cannot have all the different kinds of expertise required for the research. Eventually, they all rely on one another's testimony for connecting the various pieces of data together. When scientists accept one another's testimony, they end up having all together what none of them can have alone: the data needed to arrive at conclusions.[24]

However, philosopher Jeroen de Ridder has argued that even if scientists are expected to trust one another to give reliable testimonies, this is not always the case. Either by mistake or intentionally, scientists may not always give accurate testimonies. Therefore, something else is required, de Ridder argues, for the testimonies of scientists to be reliable and for them to be considered trustworthy. The alternative he has suggested is that scientists trust each other to make claims that are supported by evidence obtained in accordance with the prevailing methodological standards in their field. These standards can be the normative ones, that is, those that scientists officially endorse and that can be found in formal documents, such as textbooks and codes of conduct. However, these raise high expectations that do not seem to be always met in practice. But if these standards are the operative ones, that is, those that are really used in practice, then it is more realistic to claim that scientists have reasons to trust one another.[25]

Having explained what scientists are to be trusted for, it is now interesting to consider how trustworthy they are considered by lay people.

Public trust in science (and scientists)

In various polls around the world, scientists are considered to be among the most trusted professionals. For instance, a 2022 IPSOS poll across

28 countries asked participants to rate various professionals on a scale from 1 to 5, where 1 was very trustworthy and 5 was very untrustworthy. The results showed that scientists were considered as the second most trusted group of professionals, after doctors, with an average of 57% (receiving 1 or 2 on the 1–5 scale). It should be noted that only 15% on average considered scientists as untrustworthy (receiving 4 or 5 on the 1–5 scale). There of course exist variations among countries, with trust going as high as 71% in China and as low as 37% in Japan.[26] In another poll, the 2020 Wellcome Global Monitor, participants in 113 countries were asked the following two questions: "In general, would you say that you trust science a lot, some, not much, or not at all?" and "In general, would you say that you trust scientists in your country a lot, some, not much, or not at all?" Under the assumption that trust in science can only mean trust in scientists, we can perceive the first question as referring to scientists in general (e.g., all over the world) and the second as referring to the scientists in the participants' countries. Scientists were again found to be among the most highly trusted professionals, along with medical doctors and nurses. On average, those who chose 1 or 2 on the 1–5 scale were 76% when reference was made to scientists in one's country and 80% when reference was made to science in general. Again, there were variations, with science being trusted a lot by 59% of participants in Western Europe but only 22% in Sub-Saharan Africa.[27] Finally, in a 2024 survey across 28 countries, scientists were found to be trusted to do what is right by an average of about three out of four people.[28] Overall, it seems that along with medical doctors, scientists are among the most trusted professionals.

You may have noticed that even though in both the IPSOS poll and the Wellcome Global Monitor poll the answers to the questions were based on a 1–5 scale, the average results of these two polls are different. This can be explained on the basis of the different samples, the timing of the polls (before and during the COVID-19 pandemic), and other factors. Furthermore, despite the relatively high trust in scientists in both of these polls, there was a significant number of people who stated they did not trust scientists (15% in the IPSOS poll and 9% in the Wellcome poll) or that they were undecided/unsure about this (28% in the IPSOS poll and 7% in the Wellcome poll). There are several possible explanations for these results. Mistrust is more difficult to explain because one cannot know whether people have thought about their trust in science and decided that they have no strong opinion, or whether they have not thought at all about this. One could reply "neither" or "I don't know" or refuse to reply either because one is undecided or because one does not feel like having a particular view on the topic. With respect to distrust, which can be perceived as a more straightforward attitude, a likely explanation already mentioned is that people rely on previous experiences (their own or those of others) or testimonies that have caused concerns about the trustworthiness of scientists.

More generally, caution is always required in interpreting such poll results, because the responses depend on how the questions were asked, when they were asked, what options were given, and what one might mean by referring to "scientists" (e.g., government scientists vs. university researchers).[29] Indeed, a large study that distinguished among 45 science specialties has shown that what kind of scientists one has in mind may make a difference in their response. Overall, trust in scientists was found to be high, with perceptions of competence and morality having more influence on trust ratings than perceptions of assertiveness and warmth. Most importantly, whereas competence mattered equally, the impact of morality varied among the different kinds of scientists and was more strongly associated with trust in specialties working on publicly debated issues. Therefore, to document and understand trust in science, it may not be sufficient to talk about trust in science in general, but rather differentiate based on the kinds of topics scientists work on.[30] As already noted, it is policy science that is often contentious, not all science. These limitations notwithstanding, polls can indeed provide a useful overview of public attitudes and views.

The Wellcome poll included two more, quite interesting, questions: Participants were asked how much they trusted scientists to find accurate information about the world and to do their work with the intention of benefiting the public. These questions correspond to scientists' competence and intentions, respectively. More than 77% of participants globally responded that they trust the competence and the intentions of scientists a lot or somewhat. There were, however, 10% who said that they did not trust scientists entirely, and around 4% said that they did not trust scientists at all, whereas another 8%–9% were unsure or refused to reply.[31] We thus see that with respect to the key questions about scientists' competence and motivations, there seem to be more people who mistrust or distrust them, rather than altogether reject them. But what can the reasons for this distrust be?

One apparent reason for distrust in science is personal experiences of errors made in clinical settings. Unfortunately, they do occur. For instance, a study of hospital records estimated that approximately 795,000 people in the USA become permanently disabled or die annually because dangerous diseases are misdiagnosed.[32] Another study of the records of 2,428 hospitalized patients who had experienced a major clinical deterioration (either death or requiring a transfer to an intensive care unit) in the USA concluded that among the 1,863 patients who died, diagnostic error was judged to have contributed to the death of 121 of them.[33] This is an important issue that must be addressed. However, it is unlikely that this is a reason for distrusting science as a whole. Most hospitalized individuals do not die, and medical science does not represent all science—even though this is the kind of science that perhaps most people have a direct experience with.

In general, healthy skepticism and interrogation are useful for science because they motivate scientists to further confirm their conclusions or appropriately revise them based on new evidence. This stands in contrast to ill-motivated criticism, as in the case of denial. But between these two extreme cases of constructive and ill-motivated criticism of science, there have existed others where science has been questioned and eventually mistrusted. I therefore suggest that, as shown by the polls, because science distrust and mistrust likely concern many more people than denial and are definitely not due to irrational reasoning or indifference to evidence, the people who have these attitudes are the ones whom we must convince why, when, and how much science can be trusted. Furthermore, because distrust is a more active, conscious, and specifically targeted attitude than mistrust, hereafter I focus on that. However, I believe that everything that I present to counter science distrust would be equally effective to counter science mistrust.

A topic in which public distrust in science has been most evident is vaccination. This is why the present book focuses on vaccination and its history in order to gain a better understanding of science distrust and its possible causes. However, as shown in each chapter of the book, the same issues are found in other science areas too. But to get there, we need to first clarify what vaccines are and what they do.

2
Vaccination and its discontents

The pandemic that shook the world

It was in early 2020 that what came to be known as the COVID-19 pandemic (COronaVIrus Disease of 2019) began. By that time, we had all heard that the cause of the pandemic was a virus (see Box 2.1)—the so-called coronavirus. This is actually one of the many coronaviruses that exist, and it is called SARS-CoV-2. This stands for "severe acute respiratory syndrome coronavirus 2" ("2" is used to differentiate this one from SARS-CoV-1 that caused the SARS outbreak in 2002–2003). For the next two years, we had to wear masks in closed spaces, take several kinds of protective measures, and get the COVID-19 vaccines as soon as they became available. We had seen such crises in the past: SARS through the news in 2002–2003; concerns about H1N1 around 2009. But the COVID-19 pandemic came as a huge surprise, and not a good one for sure, to the western world. I was always puzzled by the view of mostly Asian people wearing masks in airports and airplanes. I never understood it before 2020. Now I do. They just knew why masks matter before I did. We nevertheless soon found out that the COVID-19 pandemic should not have been such a surprise as crucial scientific evidence already existed that such a situation might arise. In addition, even though the World Health Organization (hereafter WHO) had become aware of the situation in China very early on, the appropriate measures were never taken in a timely manner in several countries.[1]

Box 2.1 What viruses are and how they replicate

Viruses are noncellular entities and are not autonomous like cells are. This is a reason to consider viruses as nonliving. However, viruses can also enter cells and exploit them for replicating, with those cells releasing thousands or millions of new viruses as a result. This is a reason to consider viruses as living. Whether living or nonliving,

---

Trusting Science. Kostas Kampourakis, Oxford University Press. © Oxford University Press (2025).
DOI: 10.1093/oso/9780197787106.003.0002

the fact is that viruses can do harm as their replication often destroys the host cells, causing damages to the respective tissues.

Consider SARS-CoV-2 (left part of the figure below). The viral genome (an RNA molecule, similar to DNA that other viruses and all cellular organisms have) resides inside the viral particle bound to the nucleocapsid protein (N), which protects it. The resulting structure is in turn protected by a membrane that was "borrowed" from the infected cell that produced the virus. This membrane has several proteins anchored in it, such as the spike (S), the matrix (M), and the envelope (E) proteins. The S protein binds to a cell receptor and in a way "unlocks" the cell membrane, thus allowing the virus to enter the host cell. The M protein interacts with the nucleocapsid and triggers virus assembly in infected cells, for which the E protein is also required.[2]

The life cycle of a virus is very simple: They emerge from a cell and enter another cell. The molecules they consist of only serve the processes of entering cells and replicating. Here is how this process occurs (right part of the figure below). The binding of SARS-CoV-2 to the surface of a cell results in its uptake, enclosed in a viral envelope that is formed by the external membrane of the host cell. The viral genome is released and is used to produce the various proteins of the virus. New RNA molecules are also produced. The new proteins and RNA molecules assemble new virus particles (called virions), which are then released out of the host cell. This procedure is more or less the same for all viruses.

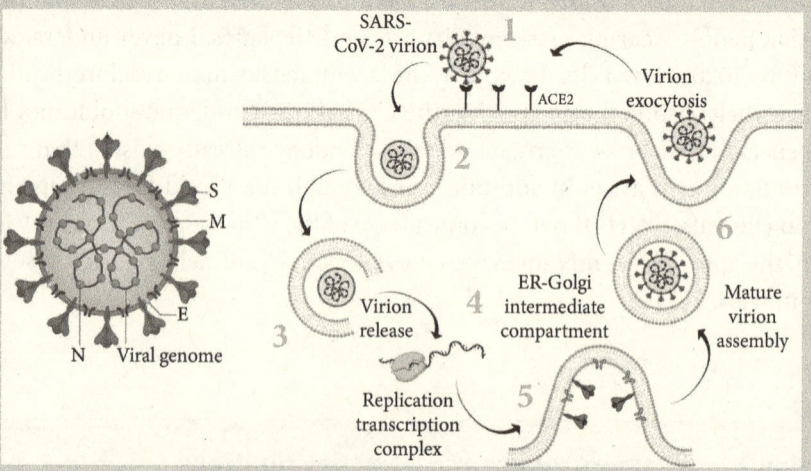

Left: The structure of SARS-CoV-2, showing the position of the viral genome as well as of the spike (S), the matrix (M), and the envelope (E) proteins. Reproduced with permission from Jackson, C. B., Farzan, M., Chen, B. *et al.* (2022). Mechanisms of SARS-CoV-2 entry into cells. *Nature Reviews Molecular Cell Biology*, 23, 3–20. *Right*: The replication of SARS-CoV-2 in cells. Its binding to the surface of a cell results in its uptake, and the viral genome is released inside the cell. New proteins and RNA molecules are produced and used to assemble new viruses out of the cell. Reproduced from Staufer, O., Gantner, G., Platzman, I. *et al.* (2022). Bottom-up assembly of viral replication cycles. *Nature Communications, 13,* 6530. (Creative Commons Attribution 4.0 International License).

The first measures that were taken at the beginning of the COVID-19 pandemic are described as nonpharmaceutical interventions (hereafter NPIs), such as lockdowns, school closures, social distancing, wearing masks in public places, and others. In theory, if we had all stayed inside our houses for two weeks, the pandemic would have been over very soon. In practice, this was simply impossible to do. It thus took more than two years for life to return to "normal"—whatever that means, because for many people life is not the same after the pandemic. Many people died; many others still suffer from problems that the viral infections caused, sometimes described as long COVID. Others lost their jobs, or saw their income diminish; others had to work from home with their young kids around because daycare or schools were closed; and all of us had to live for a long time with our social lives restricted or eliminated. The pandemic had many facets, and the political decisions about the measures that were taken had a huge impact on our life, health, and psychological well-being.

People in various countries reacted before long, resisting and protesting against the sanitary measures decided by governments. These measures included not only a variety of NPIs but also an indirect imposition of vaccination after the latter became an option for some countries in late 2020. Whereas vaccination against COVID-19 was never compulsory, there was an indirect imposition because for some time it was impossible for unvaccinated people to socialize and travel; a vaccination certificate was required. There were many concerns both about the effectiveness of the various measures as well as for the safety of vaccines. There were also concerns that all these measures violated fundamental principles such as individual liberty. All these measures were taken by governments following recommendations made by scientific bodies, such as the Centers for Disease Control in the USA and the European Medicines Agency in Europe, as well as by the scientific committees advising the various governments all over the world. The goal of these measures was to constrain the transmission of SARS-CoV-2 as much as possible. But it was evident that there were side effects that many people disliked.

Unfortunately, there was a price to pay for this. During the COVID-19 pandemic, many people died because of their reluctance to get vaccinated, sadly regretting it just before dying. In other cases, people objected to sanitary measures, often contributing to the spread of the virus with the result being that stricter measures were eventually required to control the situation and, in some cases, many more deaths than what would have been expected (which is described as the excess mortality rate). There is no doubt that mistakes were made by governments and officials in many countries at many levels, and perhaps some of these were unavoidable. But even when vaccines were produced and made available, the protests continued. The main arguments remained the same, questioning the effectiveness of the various measures and the safety of vaccines, as well as rejecting imposition and defending individual liberties.

For most people living in the western world, this seemed to be an entirely new situation, something they had not experienced before. Uncertainty prevailed. A WHO report described some of the experiences during the COVID-19 pandemic, such as school closures in Europe, as "navigating uncharted territory."[3] So did the American Enterprise Institute, a long-established and well-funded think tank in Washington, DC, when they noted that "When building closures first took effect in March 2020, many teachers, students, and parents entered uncharted waters."[4] However, neither the territory nor the waters were entirely uncharted; not everything was entirely new for those who were aware of history.

Lessons from the past

If you were taken by surprise by the pandemic, take a look at Figure 2.1. It is a photo of a group of people standing on the old station platform at Locust and Miller, in Mill Valley, California. Raymond Coyne, an Oakland resident, took this photo of his friends in November 1918, amid the so-called Spanish flu. If you look closely at the lady on the right, you will see that she carries a sign stating, "Wear a mask or go to jail." There was a mask ordinance in the fall of 1918, when the pandemic was at its first peak. On November 9, 1918, the Mill Valley Record printed an article with the headline "The Mask Ordinance—Violators Will Be Made to Pay Full Penalty," also noting that four men were arrested and fined a total of $30 for "a failure to wear the mask as required."[5] Sounds familiar? Masks and fines related to them were common in various countries during the COVID-19 pandemic. But as you can see in this photo,

Figure 2.1 A group of mask-wearing citizens, Locust Avenue, California, during the flu pandemic of 1918. Photo by Raymond Coyne. Courtesy of Lucretia Little History Room, Mill Valley Public Library. © The Annual Dipsea Race.

it was not the first time that this happened in the western world, contrary to what many of us living there had thought.

What happened back then? Between 1918 and 1920, an influenza epidemic caused the death of more than 50 million people (perhaps up to 100 million), with most deaths occurring in just a few weeks between September and December 1918. Just compare this number to the 17 million people who died during World War I of 1914–1918 (of course, in the countries directly implicated in the war, such as France, Germany, Britain, and Italy, there were more deaths due to the war than due to the flu). In fact, the misleading name "Spanish flu" is also indirectly related to the war. When the flu emerged in Spain in May 1918, it had already existed in America for two months and in France for a few weeks. However, because Spain was neutral in the war, there were no human losses there due to that, while the press reported all that happened with respect to the epidemic. So, when in June many people in Madrid fell ill within a few days, the news spread around the world. Soon everyone abroad started calling it the "Spanish flu," not being aware that it had already happened elsewhere.[6]

How did officials at the time try to address the problem? Several NPIs, such as school, theater, and church closures, public gathering bans, quarantine of suspected cases, and restricted business hours, were implemented in an attempt to restrict transmission. It has been found that the combination of fast and stringent interventions of this kind reduced peak mortality (the highest number of deaths) by half and cumulative excess mortality (mortality in excess of what was expected to happen without the pandemic) by approximately one-third on average. Therefore, there was evidence before the COVID-19 pandemic that NPIs can be effective. Furthermore, it was found that the economic disruptions during the 1918 pandemic were similar across cities, notwithstanding the measures taken. In other words, NPIs did reduce virus transmission during the 1918 pandemic without further affecting the economy.[7]

Figure 2.2 shows the weekly mortality rates for both influenza and pneumonia, as well as the mortality acceleration date and the start and end dates of three different types of NPIs (public gathering bans, school closures, and quarantine/isolation) for six different cities. The association between the NPIs and mortality rates is evident. You do not need to look at the details of the graphs, but only grasp the overall pattern: The longer the duration of the measures taken, represented by the length of the horizontal bars, the lower the death rate, represented by the height of the curves. For instance, Boston and Philadelphia experienced increases in mortality earlier and were slower to implement measures than Rochester or St. Louis. The latter is an example of the quick and effective implementation of social distancing measures to manage the pandemic, especially compared to the mortality rates in Boston and Philadelphia. In fact, medical doctors in St. Louis believed that the timely implementation of measures had prevented the high mortality experienced by the two eastern cities. Another interesting comparison is that between the Twin Cities of Minneapolis and St. Paul. Although the flu emerged there around the same time, officials in Minneapolis moved quickly to ban public gatherings and close schools in early October, whereas St. Paul remained largely open into November as its officials believed that closures would not be effective.[8]

I must note again that I do not mean to suggest that NPIs always work without any side effects, or that the measures taken during the COVID-19 pandemic were appropriate in all cases. Things are never black and white, and there is no single solution that would work perfectly everywhere. All I want to show is that there was evidence before the COVID-19 pandemic that NPIs could work. I am inclined to think that it is on the basis of this kind of evidence that scientists made their recommendations. But were people made

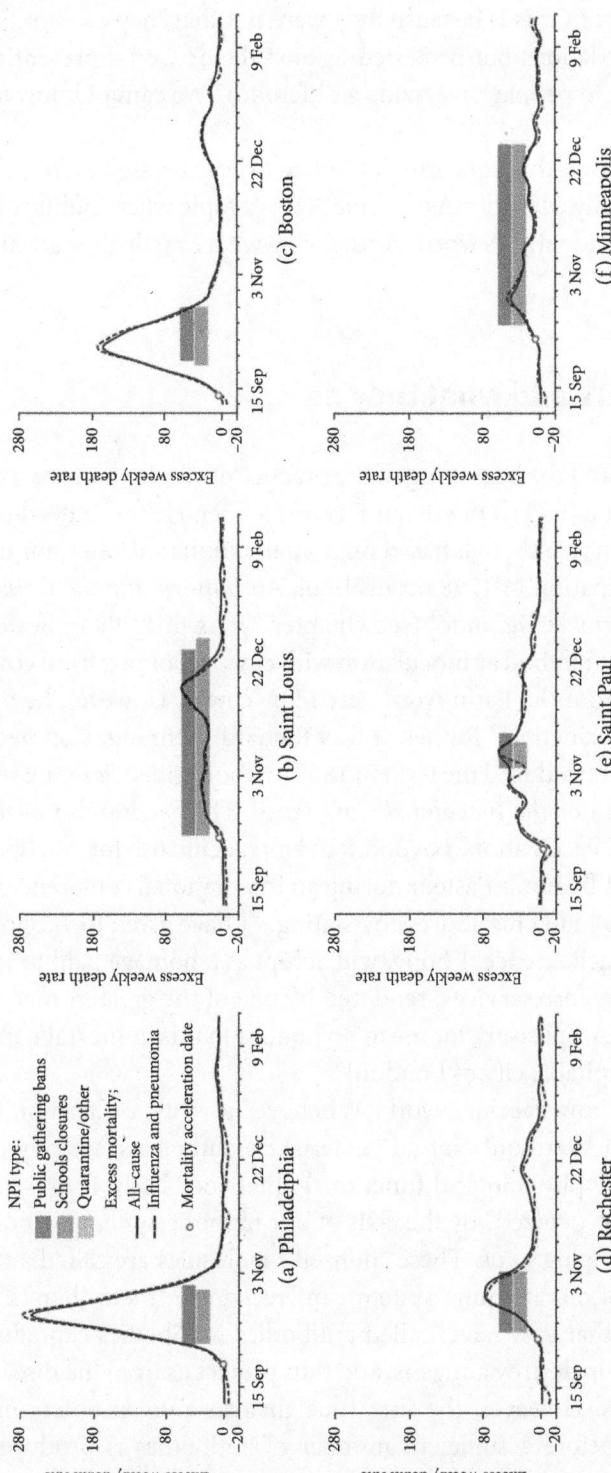

Figure 2.2 The impact of nonpharmaceutical interventions such as school, theater, and church closures, public gathering bans, quarantine of suspected cases, and restricted business hours in several US cities during the 1918 pandemic. Overall, the longer the duration of the measures taken (represented by the length of the horizontal gray bars in the graphs), the lower the death rate was (represented by the height of the curves in the graphs). (Reprinted with permission from Figure 3 in Correia, S., Luck, S., & Verner, E. (2022). Pandemics depress the economy, public health interventions do not: Evidence from the 1918 flu. *The Journal of Economic History, 82*(4), 917–957).

aware of this evidence? Or is it because they were not that they did not follow these recommendations but protested against them? Could presenting this kind of evidence to people have made a difference? We cannot know for sure.

But what we do know is that concerns continued to be expressed even when vaccines became readily available. As vaccination is a topic where public distrust is most clearly evident, it is worth considering what exactly they are and what they do.

What vaccines are and what they do

The idea of vaccines and the beginning of the practice of vaccination are usually associated with the English physician Edward Jenner. He was indeed the first to provide evidence, although based on a small number of cases, for the effectiveness of vaccination in 1798, in his book *An Inquiry into the Causes and Effects of the Variolae Vaccinae*[9] (see Chapter 3). As this title indicates, Jenner referred to that method as inoculation with cowpox, or pox from cows (*Variolae Vaccinae*, from the Latin word *vacca* for "cow"). However, he did not use the term "vaccination." Rather, it was Richard Dunning, a surgeon from Plymouth, who introduced the term in 1800 in a book titled *Some Observations on Vaccination or the Inoculated Cow-Pox*.[10] The general use of the terms "vaccine" and "vaccination" beyond Jenner's specific use for smallpox was proposed in 1881 by Louis Pasteur during an international conference in London. Pasteur concluded his address by stating: "I have given to vaccination an extension which science, I hope, will accept as a homage paid to the merit, and to the immense services, rendered by one of the greatest men of England, Jenner. What a pleasure for me to do honour to this immortal name in this noble and hospitable city of London!"[11]

Let us see briefly how vaccines work. Whenever a virus, bacterium, or any other pathogen (Attention! Not all bacteria are pathogens; we rely on many bacteria for our physiological functions) enters our body, it presents molecules that are "recognized" by the cells of our immune system as "nonself," that is, not belonging to us. These "nonself" molecules are called antigens, and the cells of our immune system can "recognize" them thanks to specialized proteins that they have, called antibodies. Antibodies can effectively neutralize or help destroy antigens, and thus protect us from the disease caused by pathogens. However, the first time that an antigen enters our body, it takes time before a sufficient number of antibodies is produced.

This happens because a process of cell recognition and activation must first take place. In some sense, the cells of our immune system have to "identify" the antigen, before producing the respective antibodies in sufficient amount. Therefore, the first time we are exposed to an antigen, the production of antibodies is slow (this is called the primary immune response). However, this initial exposure also results in the production of memory cells, which will "remember" and immediately "recognize" the respective antigen the second time it enters our body. Thus, the secondary immune response (the second and any subsequent time we encounter this antigen) is much faster and much more effective (Figure 2.3).

This is how vaccines confer protection. Instead of waiting when we are going to encounter a pathogen and risk developing the disease, we introduce its antigens in a harmless form in our body. This can be done in various ways, and so there exist different vaccine types (Box 2.2). In all these cases, the antigens of the pathogen can be "recognized" by our immune

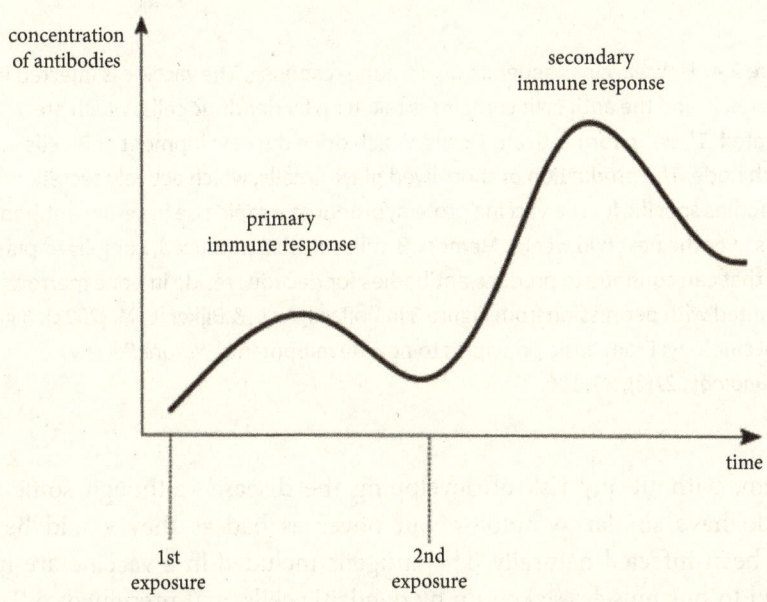

Figure 2.3 Vaccination activates our primary immune response. As a result, if we are infected by the respective pathogen, our immune system will be ready to initiate a secondary response, which will have a higher antibody production and which will thus likely be protective. In short, thanks to vaccines, the first time we are infected, our immune system reacts as if it were the second one.

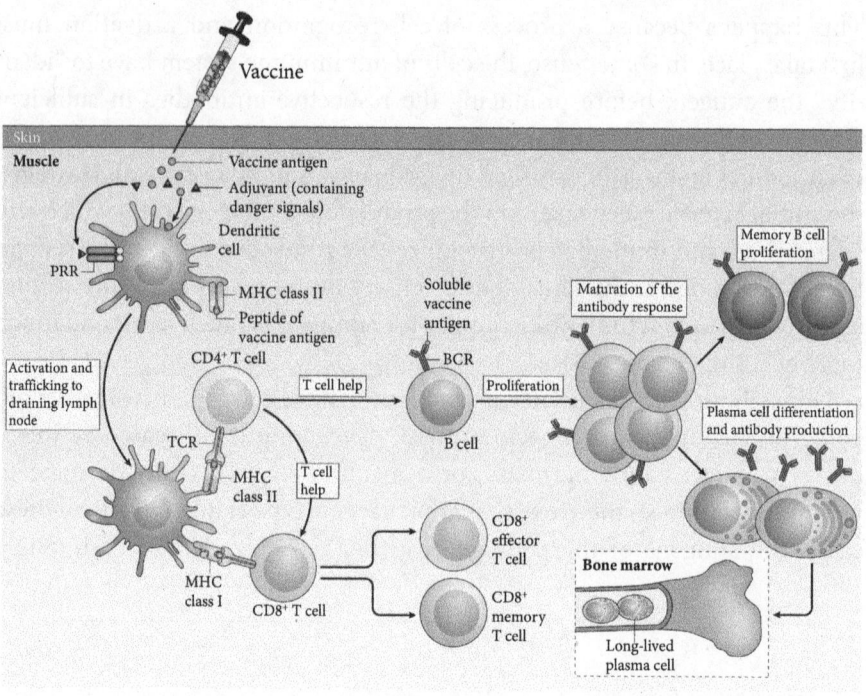

Figure 2.4 How vaccines generate an immune response. The vaccine is injected into the muscle and the antigen it contains is taken up by dendritic cells, which are activated. These in turn activate T-cells, which drive the development of B-cells in the lymph node. The production of short-lived plasma cells, which actively secrete antibodies specific for the vaccine protein, produces a rapid rise in serum antibody levels over the next two weeks. Memory B-cells are also produced. Long-lived plasma cells that can continue to produce antibodies for decades reside in bone marrow. Reprinted with permission from Figure 3 in Pollard, A. J., & Bijker, E. M. (2021). A guide to vaccinology: From basic principles to new developments. *Nature Reviews Immunology, 21*(2), 83–100.

system, without any risk of developing the disease—although some people do have similar symptoms, but never as bad as they would be had they been infected naturally. The antigens included in a vaccine are introduced to our muscles, taken up by dendritic cells, and presented to T-cells (white blood cells, also called T-lymphocytes), which in turn activate the B-cells (another type of white blood cells, also called B-lymphocytes). This results in the production of antibodies over two weeks, as well as of memory cells (Figure 2.4). As a result, when the pathogen enters our body for the first time and we are already vaccinated, our T-cells and B-cells will react as if it were the second; in other words, there will be a secondary immune response the first time we are infected, and thus no disease. This is also why we may get a disease once (the first time we are infected), but not again (because the secondary immune response protects us).

Box 2.2 Types of vaccines

There exist several types of vaccines, which differ in the way they generate an immune response. The vaccines against SARS-CoV-2 can be divided into three broad categories:

- Protein-based vaccines that produce antigens outside host cells (production of molecules outside cells is often described as in vitro), such as:
  - *Inactivated virus vaccines*: Viruses are physically or chemically inactivated but preserve the integrity of the virus particle, which serves as the antigen.
  - *Virus-like particle vaccines*: Viral proteins are produced to form noninfectious viral particles, which lack the viral genome.
  - *Protein subunit vaccines*: Only key viral proteins are included in the vaccine, such as the spike (S), the matrix (M), and the envelope (E) proteins.
- Gene-based vaccines that deliver genes (DNA sequences encoding proteins) of viral antigens to host cells wherein the antigens are produced (production of molecules within cells is often described as in vivo), such as:
  - *Virus-vectored vaccines*: Genes encoding viral protein antigens are included into virus vectors (such as adenovirus). These enter cells in which the antigens are produced.
  - *DNA vaccines*: Viral antigens are produced within host cells by the DNA included in the vaccine. DNA produces mRNA, which in turn produces the protein antigen.
  - *mRNA vaccines*: Viral antigens are produced within host cells by the mRNA included in the vaccine. The mRNA directly produces the protein antigen.
- A combination of both protein-based and gene-based approaches to produce protein antigen or antigens both in vitro and in vivo, which typically include:
  - *Live-attenuated virus vaccines*: Viruses are attenuated with different methods, becoming nonpathogenic or weakly pathogenic but retaining the ability to generate an immune response by resembling the infection by a live virus.

Among these types, DNA and mRNA vaccines have the advantage of rapid manufacturing against emerging pathogens.[12]

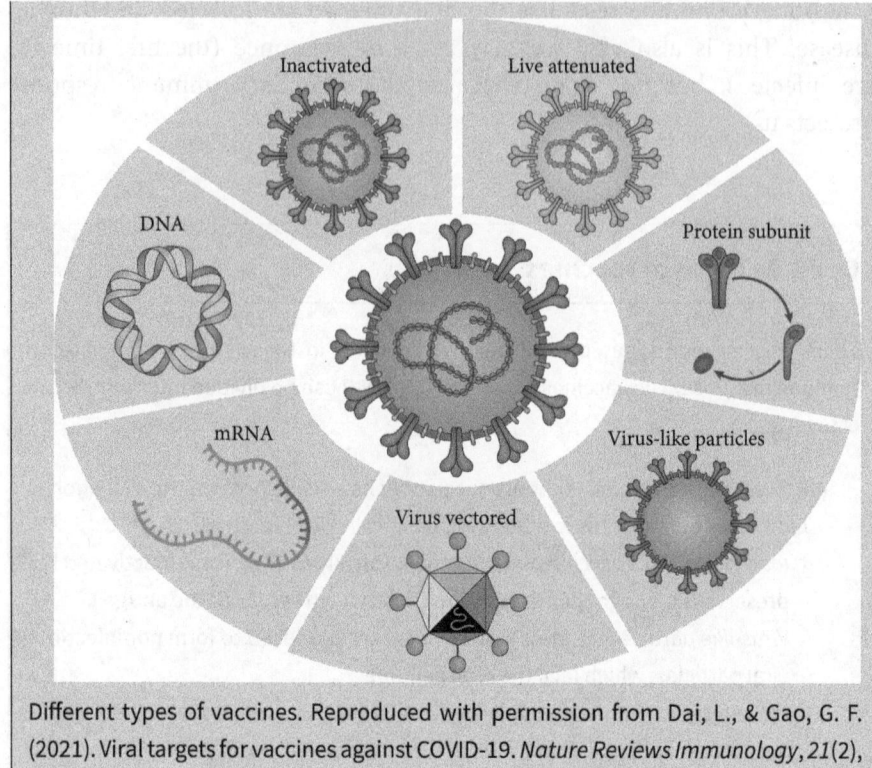

Different types of vaccines. Reproduced with permission from Dai, L., & Gao, G. F. (2021). Viral targets for vaccines against COVID-19. *Nature Reviews Immunology*, *21*(2), 73–82.

The main goal of vaccination is therefore preventative, by inducing the immune response against a pathogen before we encounter it. Depending on the measures taken, vaccination can bring about one of the following outcomes:

- *Control:* The reduction of disease
  - incidence (the number of new cases of disease in a population)
  - prevalence (the proportion of a population with a specific disease)
  - morbidity (the disease state of a population)
  - mortality (the number of people who die from the disease)
  to a locally acceptable level because of deliberate efforts; continued intervention measures are required to maintain this reduction.
- *Elimination of disease*: Reduction to zero of the incidence of a disease in a defined geographical area as a result of deliberate efforts; continued intervention measures are required.
- *Elimination of infections*: Reduction to zero of the incidence of infection caused by a specific agent in a defined geographical area as a result of

deliberate efforts; continued measures to prevent the reestablishment of transmission are required.

- *Eradication*: Permanent reduction to zero of the worldwide incidence of infection caused by a specific agent as a result of deliberate efforts; intervention measures are no longer needed.
- *Extinction*: The specific infectious agent no longer exists in nature or in the laboratory.[13]

As it is individuals who are vaccinated, the immediate outcome of vaccination is individual immunity and protection. However, because vaccines never provide 100% protection and because some people cannot be vaccinated (e.g., because they have compromised immune systems or because they are too young), individual immunity is not attainable by all. But these people can nevertheless be protected if communities reach what is described as collective, or herd, immunity. Simply put, the more people are vaccinated, the less likely it is for the pathogen to be transmitted and cause disease. On the one hand, vaccinated people are unlikely to have severe symptoms, and the disease will last less than usual, thus constraining the reproduction and the transmission of the pathogen. On the other hand, people who for whatever medical reason cannot be vaccinated are protected thanks to the protective barrier formed by the vaccinated ones. It is therefore important to keep in mind the impact of vaccines at both the individual and the community level. The more people are vaccinated in a population, the less likely it is for the pathogen to propagate, and thus the more likely herd immunity is to be reached (Figure 2.5). The threshold required for herd immunity depends on the contagiousness of the pathogen. For highly contagious viruses such as the one causing measles, more than 95% of the population needs to be vaccinated in order for herd immunity to be reached.

Achieving herd immunity, in turn, requires vaccines that work well, which is described by the terms "vaccine efficacy" and "vaccine efficiency." Even though these terms are both roughly about how well vaccines work, they are not synonymous. Before a vaccine is made available, it is tested in controlled clinical trials, usually randomized and double-blind. What these terms mean is that a certain number of people volunteer to be randomly assigned to two groups, without knowing to which one they are assigned: One group consists of people who will receive the vaccine, and the other of people who will receive a placebo (a solution containing everything else, except the antigens). Depending on how many people in each group develop the disease, we can estimate the efficacy of the vaccine. Take a look at Figure 2.6. Imagine that in the placebo group (top), 100 people developed the disease compared to 20 people in the vaccine group (bottom). This means that compared to 100 people in the placebo group who developed the disease, there were 80 fewer people (100 – 20 = 80) in the vaccine group who did not develop it. When the people who did not develop the disease in the vaccine group are 80% fewer than the placebo group, we say that the vaccine has an efficacy of 80%. An efficacy of 80% does not mean that the vaccine works only in 80% of the cases! It is about an 80% reduction in the number of cases.

Once the results of the clinical trials are considered satisfactory, the vaccine is made available to the general population. But as the people participating in the clinical trial are not necessarily representative of the whole population, we cannot know in advance whether the vaccine will work well in real-life conditions. This is described as vaccine effectiveness, which is generally measured by comparing the frequency of health outcomes in vaccinated and unvaccinated people. As the conditions in the clinical trial may differ from the real-world conditions, it is possible that vaccine effectiveness differs from

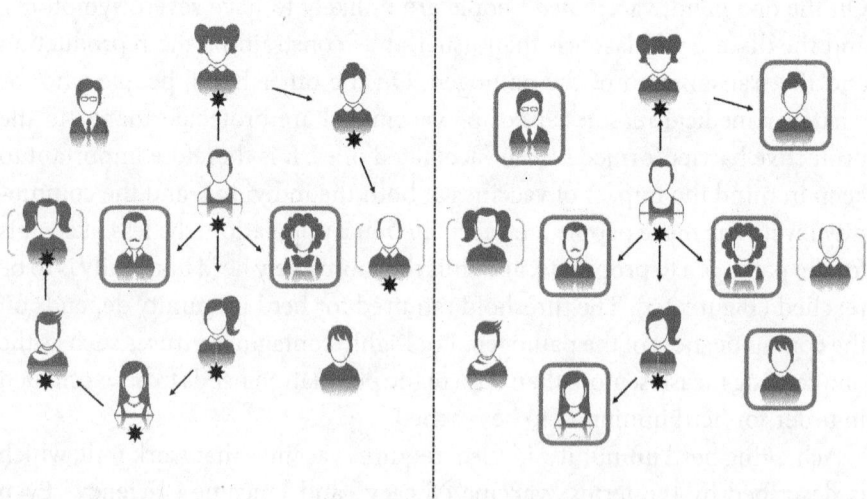

Figure 2.5 *Left*: When unvaccinated individuals are in contact with ones infected by the pathogen (represented by an asterisk), its transmission is possible. The vaccinated, and therefore immune, individuals (within rectangles) are not sufficient to prevent this transmission. This exposes those who cannot be vaccinated for medical reasons (within brackets) to danger. *Right*: The more vaccinated individuals there are in the population, the more difficult it is for the pathogen to spread and affect the susceptible ones (see http://rocs.hu-berlin.de/D3/herd/ for a simulation). This is how vaccination can help achieve both individual and herd immunity, and eventually confer protection even for the unvaccinated individuals. Image credit: "Freepik.com." This figure has been designed using assets from Freepik.com

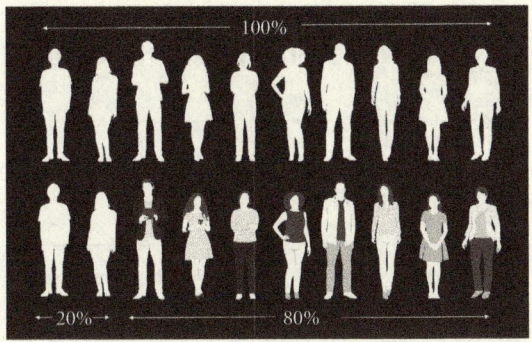

Figure 2.6 If in a placebo group (*top*) 100 people develop the disease, and in the vaccinated group these are only 20, we can say that because the people who did not develop the disease in the vaccine group are 80% (100 – 20 = 80) less than the placebo group, the vaccine has an efficacy of 80%. A vaccine efficacy of 80% means that 80% fewer people will contract the disease when they come in contact with the virus, not that it will only work in 80% of cases. Image credit: "Freepik.com." This figure has been designed using assets from Freepik.com

the vaccine efficacy measured in a trial. A recent analysis[14] of the available data until December 2022 has shown that vaccine effectiveness against SARS-CoV-2 infections, hospitalizations, and mortality generally decreases over time. Whereas it was initially 92% for hospitalizations and 91% for mortality, it reduced to 79% at 224–251 days for hospitalizations and 86% at 168–195 days for mortality.

It would have therefore been expected that once the high efficacy of the COVID-19 vaccines (Box 2.3) had been established, people would rush to get vaccinated in order to diminish the risk of disease, and hopefully return to a life without restrictions. But this was not the case at all. Many people were reluctant to get vaccinated, a situation that is described with two terms that are often confused: *vaccine refusal* and *vaccine hesitancy*. Vaccine refusal is a behavior, as is vaccine acceptance; people may decide to receive a vaccine (acceptance), or they may decide not to do so (refusal). These two behaviors are different from vaccine hesitancy, which is an attitude characterized by ambivalence regarding vaccines; vaccine hesitancy can actually comprise a range of attitudes from mild to severe uncertainty about the safety and the effectiveness of a vaccine and—depending on the situation—may lead to either vaccine acceptance or vaccine refusal.[15] Vaccine hesitancy is directly related to trust. Vaccine hesitant people are those who mistrust or distrust vaccines. It is therefore interesting to consider how many vaccine hesitant people exist.

Box 2.3 Efficacy of COVID-19 vaccines at the time of their approval

It should be noted that the efficacy of these vaccines cannot be directly compared because of differing clinical trial designs.[16]

Manufacturers (vaccine name)	Type (see Box 2.2)	Doses	Efficacy against symptomatic disease (percentage)
Pfizer and BioNTech (Comirnaty)	mRNA	2	95
Oxford and AstraZeneca (AZD1222)	Viral vector	2	82.4
Moderna and NIH (mRNA-1273)	mRNA	2	94.5
Gamaleya (Sputnik V)	Viral vector	2	91.6
CanSinoBio (Convidecia)	Viral vector	1	65.7
Novavax (NVX-CoV2373)	Protein	2	95.6
Johnson & Johnson (Ad26.COV2.S)	Viral vector	1	72
Sinopharm (BBIBP-CorV)	Inactivated virus	2	79.34
Sinovac (CoronaVac)	Inactivated virus	2	50.4
Bharat Biotech (Covaxin)	Inactivated virus	2	Unknown

The 2018 Wellcome Global Monitor, a survey conducted in 126 countries, included a set of questions about vaccines in an attempt to assess whether trust in science is related to vaccine confidence. Two of these questions were about whether participants agreed or not with the following statements: "Vaccines are safe" and "Vaccines are effective." Overall, approximately eight in ten people agreed that vaccines are safe and effective. Most of the other participants were unsure or did not know, but there were some (less than one in ten) who disagreed with both vaccine safety and effectiveness.[17] What about the COVID-19 vaccines? One might reasonably expect that in contrast to vaccines that have been administered to millions of children, people would be more skeptical of the newly produced COVID-19 vaccines. However, a study in 23 countries that aimed at investigating COVID-19 vaccine hesitancy in June 2021 found that approximately three in four people had received at least one dose of a COVID-19 vaccine or were willing to receive it when it became available (this was defined as COVID-19 vaccine acceptance). In the same study, the researchers also looked into vaccine hesitancy, which they defined as having reported "no" to the question about whether people had received at least one dose of a COVID-19 vaccine and also either "unsure/no opinion," "somewhat disagree," or "strongly disagree" to the question on whether they would take a COVID-19 vaccine when it was made available to them. This categorization is problematic because it lumps together vaccine refusal with vaccine hesitancy, even though the latter does not necessarily lead to the former. This notwithstanding, one in four people were found to be skeptical about the COVID-19 vaccines.[18]

One might think here that there is no big problem as the majority of people trust vaccines. However, when there are infectious diseases that require more than 90% of the population to be vaccinated in order for herd immunity to be achieved, the current state of vaccine acceptance is not sufficient. Given the scientific consensus that the vaccines approved by the relevant authorities are both safe and effective, it is notable that about one in five people do not trust this consensus. But why?

Well-known causes of vaccine and science distrust

In her exhaustive study of vaccine hesitancy, philosopher Maya Goldenberg has correctly identified three main causes of science distrust: (1) racism, discrimination, and injustice on behalf of scientists; (2) financial conflicts of interest in health research and practice; and (3) social media. She noted that a lot more attention has been given to social media than to the other two causes. But even though acquiring unreliable information through social media may seem like a main cause of science distrust, it may also be the effect of the other two causes. Goldenberg argued that the perception that science favors discrimination and that it is guided by vested interests may drive those who think so away from the official sources of information, thus making them prone to accept the unreliable information they find on social media. Thus, Goldenberg suggested, more attention should be given to science-related racism, discrimination, and injustice, as well as to clarifying the relation between pharma and science, in order to understand why social media becomes for many a source of information instead of the official sources.[19] Let us consider these two causes in more detail.

Physician and science writer Ben Goldacre has analyzed in detail the problems in the scientific research done or funded by pharmaceutical companies. In a book-length account, he has criticized the problems of industry-sponsored studies that are more likely to produce results in favor of the sponsor's product. Simply put, Goldacre argued, when companies pay for research on their products, they prefer that it shows that the products are safe and efficient, rather than the opposite. Companies also spend as much on research as they spend on their marketing activities for obvious reasons—they want to sell their products. This can in turn influence the decisions of regulators and physicians among others.[20] Goldacre's suggestion about the remedy for the problem is a well-operating system of medical research for the benefit of all. However, I am inclined to think that discovering cases of fraud or misconduct on behalf of pharmaceutical companies should not be surprising. As with any other commercial product, manufacturers will unfortunately exist who will use deceit to sell it. This is neither the norm nor good, of course; but it is not a surprise either as it happens in many commercial activities. The pharmaceutical industry has no reason to differ from other industries when it comes to profit. Of course, as with all industries, it should follow all the relevant regulations, with respect to controls and transparency.

Pharmaceutical companies have been involved, and have also involved scientists, in fraud or other not-so-legitimate or not-so-legal activities. Discussing these cases in detail falls outside the scope of the present book, but biotechnology consultant Stewart Lyman has produced an interesting list of such cases. What makes this list interesting is not only the companies involved but also the kinds of practices implemented. These include raising prices unjustifiably, using deceptive marketing practices, faking or falsifying data, failing to warn about serious risks on product labels, behaving unethically, profiting in not-so-legal ways, bribing, or failing to disclose required clinical trial results. Perhaps there is more. Lyman also cited the report of a nonprofit consumer advocacy organization about the major financial settlements and court judgments between pharmaceutical companies and federal and state governments from 1991 through 2015. According to that report, the companies entered into 373 settlements for a total of $35.7 billion in criminal and civil penalties.[21] Given all these, according to Lyman, it should be no surprise that people are suspicious of pharmaceutical companies and the various science-related financial interests.[22] As it is obvious why this is a cause of distrust in science, I have decided not to deal with it in detail in the present book.

There is however a not-so-obvious aspect of all this I want to consider: how sensitive and suspicious people are when it comes to pharmaceutical companies, to the extent that we may forget how many lives are saved every year by their products. As historian Andrea Grignolio has correctly pointed out, people seem to forget that anything that we use in health treatment, most importantly drugs and vaccines, is produced by pharmaceutical companies, and is more important to our survival than our smartphones, computers, or cars. Yet, it is unlikely that we will see many people praising pharmaceutical companies for what they offer. Scandals and corruption exist, and whenever they occur, the people responsible should be brought to justice. But we should also remember how much pharmaceutical companies offer to our everyday lives.[23] For instance, Pfizer may have made almost $57 billion from their COVID-19 vaccine and drug in 2022.[24] This is money that should be declared in a transparent manner and taxed properly. But we should also think beyond the money, about how many lives were protected by their vaccine and drug (as well as by those of other companies), and also what might have happened if Pfizer and the other companies were not prepared to conduct the expedited process of COVID-19 vaccine production. In short, with pharmaceutical companies, as with every commercial company, we should both criticize the bad aspects and acknowledge the good ones.

Social media is another well-known reason for distrust in science. I agree with Goldenberg's assessment that this is not only a cause but also an effect of other causes of distrust. We all know that social media is about the transmission of information, and so there can be questions about its accuracy, as well as about the validity and reliability of the respective sources. But a very important distinction exists between misinformation and disinformation, which is essentially the distinction between misunderstandings and lies, between common misconceptions and targeted manipulation.[25] When it comes to science, both are at play. Misinformation and disinformation about vaccines have been a well-studied cause of distrust. For instance, the tweets of Donald Trump about vaccination—all of them before he became President of the USA—have had a documented impact on people. In particular, it has been found that Trump's voters were more concerned about vaccines than other people in the USA, and their reading of Trump's anti-vax tweets increased vaccination concerns among them.[26] This is why media literacy should be part of the educational curriculum at as early an age as possible. Students should learn from very early on to evaluate the reliability and the validity of various sources of information. Teachers should discuss with their students news articles about science: their purpose, how and when they were written, what the author's claims are, and so on. But this is an issue that is not specific to science.[27] For this reason, I have decided not to deal with this in detail either in the present book.

Therefore, I will only consider in detail the first of Goldenberg's known causes: discrimination and injustice. The reason for this is that it comes from within science, like the other causes I consider in the present book. While I agree with Goldenberg about the importance of trust in science, I disagree with her view that questioning science is not due to ignorance of science or misunderstandings of science.[28] Goldenberg is right that this is not all that matters; she is also right if by "science" she refers to science content knowledge. Most importantly, she is right that scientists and authorities are quick to blame the ignorant public. In fact, I think that scientists and authorities are themselves to blame for science distrust not only because they do not invest the necessary time to listen to the concerns of lay people but also because they have not paid attention to the kind of education that lay people need in order to be able to understand the multiple issues around vaccination.

In the present book, I argue that misunderstandings of certain aspects of science other than content knowledge are causes of lay people's lack of trust in science. In fact, the present book is devoted to uncovering these mostly underexplored causes of science distrust.

Some underexplored causes of science distrust

A close look at the history of vaccination reveals what I consider some overlooked causes of science distrust: a lack of understanding of how science is done, and how it relates to issues at its interface with society—what in the science education literature is described as the nature of science.[29] I argue that lay people may distrust science because they do not understand its methods, assumptions, and limitations, as well as because they may not realize what questions science can and cannot answer.

One key issue with vaccines is that whereas we can measure and establish their effectiveness, it is less easy—at least initially—to establish their safety. But when particular vaccines have been used for decades in millions of people, their safety can be legitimately established. Vaccines, like all medical interventions, can entail some risk, which should always be compared to the risk of developing the disease that a vaccine is intended to protect us from. But people often overestimate the risk of vaccine side effects, even if these are very rare, and underestimate the risks of infection itself. For many parents, the idea of introducing a part of a pathogen in the body of their children just for the purpose of prevention is disturbing. Many parents can accept the potential side effects of drugs because they are administered to counter a disease that is already there. But administering a vaccine with potential side effects, albeit extremely rare, to counter a disease that is not there is something that is much harder to accept. For this reason, to be able to deal with these risks, it is necessary for lay people to understand that risks and uncertainty are inherent in science.

Therefore, understanding the nature of science must become a central learning goal of school teaching. But more urgently, we need to educate the citizens of today who most likely were not taught about the nature of science during their school years. If you are one of them, then the present book aims to provide you with opportunities for the kind of reflection about science you may have not been given during your school years. This is a prerequisite not only for improving the present but also for building the future. Only if people realize that the future generations, including their own children, would benefit from a different kind of science teaching than they themselves experienced in school could a reconceptualization and reform of science teaching such as the one suggested in the present book be possible. Changing science teaching that has been content-focused for generations could take place only if it were supported by society.

The question then becomes: Which aspects of the nature of science do lay people need to understand? Besides the already known causes of discrimination and injustice on behalf of scientists, my own close study of the history of vaccination has revealed seven other potential causes of science distrust, and these are the ones I consider in the subsequent chapters. Therefore, I suggest that lay people may distrust science because:

1. of discrimination, injustice, and potential conspiracies, on behalf of scientists;
2. of the perceived implications of scientific conclusions for their worldviews;
3. they dislike the uncertainty that is inherent in science;
4. of their lack of understanding of what constitutes scientific evidence;
5. they have concerns about the reliability of the procedures internal to science;

6. of the conflicting claims of scientists and other "experts";
7. they confuse the scientific conclusions with the related political decisions; and
8. of the lack of immediately observable evidence for the claims of scientists.

Each of these causes is considered in a respective chapter of the present book
(Chapters 3–10). In each of these chapters, an episode from the history of
vaccination that reveals a particular cause of distrust is first considered. Then
the key aspects are clarified, and they are also considered in the context of
science domains other than vaccination. Finally, the reader is guided to reflect
upon these aspects in order to better understand them.

   Let us then begin exploring these underexplored potential causes of science
distrust.

3

"After they give us the blood tests ... they said we had bad blood"

Scientists' discrimination, injustice, and ... conspiracies

"Bad blood"

> Anyhow, they came around give us the blood tests. After they give us the blood tests, all up there in the community, they said we had bad blood. After then they started giving us the shots and give us the shots for a good long time. I don't remember how long it was. But they got through giving us those shots, they give me a spinal tap.[1]

This is Charlie Pollard, testifying in a Congressional hearing convened by Senator Edward Kennedy in 1973.[2] On July 26, 1972, reporter Jean Heller of the Associated Press wrote an article for *The New York Times*, titled "Syphilis Victims in U.S. study went untreated for 40 years." Heller reported that "human beings with syphilis ... were induced to serve as guinea pigs," being left untreated for the disease. The reason? "... to determine from autopsies what the disease does to the human body." Who was responsible? The US Public Health Service (PHS). The Assistant Secretary of Health, Education and Welfare for Health and Scientific Affairs was said to have "... expressed shock in learning of the study." Dr. J. D. Millar, who was head of the CDC division overseeing the study, noted that a "serious moral problem arose," because even though in the late 1940s penicillin had been shown to be an effective cure for syphilis in its early stages, it was nevertheless not administered to the study participants. As a result, several of them died of syphilis even though their deaths could have been avoided. For many people, this was a shocking revelation of the potential for scientific abuse, especially given that all participants in the study were Black people.[3]

Syphilis, also called the Great Pox, is a highly contagious disease, caused by the bacterium *Treponema pallidum*. Infection can occur either through body contact, usually sexual intercourse, or from an infected mother to her fetus

Trusting Science. Kostas Kampourakis, Oxford University Press. © Oxford University Press (2025).
DOI: 10.1093/oso/9780197787106.003.0003

during pregnancy. Due to many practical difficulties, there exists no vaccine for this disease yet. Even though in practice there was no race-based differentiation in the treatments that people received, it was widely thought among (White) officials and physicians that syphilis occurred differently and resulted in different pathological conditions in White and Black people. Based on their personal observations, rather than epidemiological data, they attributed the occurrence of syphilis in Black people to their sexual behavior, ignoring the numerous cases of infants born with the disease. They also thought that Black people did not take the disease seriously, and that they were not willing to do anything about it. Thus, these (White) officials and physicians believed that syphilis in Blacks was inevitable.

An influential study by Ernest Zimmerman published in 1921, "... undertaken for the purpose of emphasizing inherited racial differences in response to syphilitic infection," arrived at the conclusion that there existed "... certain clinical characteristics of syphilis in the negro which can hardly be explained on grounds other than inherited biologic differences."[4] This was a study of 1,843 patients with syphilis, in the charity clinic of Johns Hopkins University. Zimmerman reported that many more Black than White people had developed syphilis before their 20 years of age. Confirming previous views, he blamed Black people themselves for the disease, by suggesting: "That over 50%, of colored female syphilitics have been infected before the twenty-first year is due chiefly to their early indulgence in promiscuous sexual intercourse, the greater frequency of syphilis in that race, and the failure of the colored male with primary and secondary lesions to abstain from promiscuous intercourse."[5] Most importantly, Zimmerman concluded that there were "inherited biological differences" in how syphilis developed. Blacks were more likely to experience reactions on their skin and bones, and less likely to experience neurological problems. In contrast, in Whites, there were fewer or no noticeable skin symptoms, but there was a higher chance of developing specific nervous system problems later on. For many, the findings of Zimmerman's study were indisputable.

What was really indisputable at the time, however, was that Black people were on average less privileged, lived under worse sanitary conditions, and had less access to healthcare than Whites. It was soon realized that social status rather than race was a more important health determinant. Yet racial prejudices persisted. Booker T. Washington, the founder of the Tuskegee Institute, thus drew the attention of Julius Rosenwald and his philanthropic fund to help ameliorate the living conditions of Black people in the US South. When Michael Davis was appointed director of the medical services of the Rosenwald Fund in 1928, he tried to improve the living conditions of Black people by encouraging the training and employment of Black nurses, by enhancing the training of Black physicians, and by establishing racially mixed hospitals. But the implementation of all these measures required the support of the PHS, and the person appointed to this role was Taliaferro Clark. One of the proposals approved by the Fund for 1930 was about the control of venereal disease in the rural South. One of the places chosen to do this was Macon County in central Alabama, with a population of just over 27,000 people in 1930, 82% of whom were Black.

The living conditions in Macon County were terrible. People suffered from chronic malnutrition, and many lived in houses without indoor plumbing. Access to healthcare was almost nonexistent, and illiteracy accompanied these conditions, despite the presence of the Tuskegee Institute. Among 16 private physicians practicing medicine in Macon County in the early 1930s, only one was Black, and most people could not afford their services. As a result, most Black people consulted physicians only in emergencies, and thus chronic diseases such as syphilis were left untreated. So, when a syphilis control program was announced in 1930 for Macon County, health officials responded enthusiastically. The initial plan was to test people widely for syphilis and then implement a one-year treatment program. As most Black people worked on plantations, the cooperation of their White bosses was necessary. Convinced that the diagnosis and treatment of syphilis in their workers was to their own advantage, the White planters not only allowed but also sometimes ordered their workers to participate in the syphilis program. This happened without explaining to participants what was happening, and without asking for their consent. Worse, Black people obeyed these orders and cooperated willingly with the program, because obeying White rule was all they had known during their lives. Above all, Black people were eager to get some medical attention, as this was something they normally were not able to afford on their own.

Dealing with people who had never before in their lives been treated by a doctor was difficult for public health officials. Thinking that there was no point in trying to explain to participants what they were doing, and that they would better communicate with them by speaking the rural Black argot, they announced that they had come to test people for "bad blood." Unfortunately, this meant different things to different people, and caused confusion among participants that health officials ignored. As a result, Black people in Macon County thought that they were being tested and treated for whatever disease they had. Worse, they were not made to understand what the specific disease they were tested for was, and that it was contagious. The health officials simply wanted to get their job done, apparently thinking that there was no point in trying to conduct social hygiene work with poorly educated people. At the same time, the findings were alarming. Macon County had the highest rate of syphilis, a shocking 36%, which could easily reinforce the stereotype of syphilis being a disease of Black people.

The Rosenwald Fund was monitoring the program closely. The first inspector to be appointed was Dr. H. L. Harris, a Black physician who had not grown up in the South and who was thus shocked by the living conditions he encountered in Macon County. He was concerned that the treatment might not be administered properly because of the working conditions, the exhausted medical staff, and the lack of thorough physical examinations. In the end, he concluded that the community needed a comprehensive health and social welfare program. Therefore, he recommended that the program should not be extended as it had accomplished what could have been hoped from it. Despite disagreements, and as the Great Depression put pressure on the Fund as the stock market had declined, the trustees decided not to further support the syphilis program. But not everyone was ready to let it go.

Taliaferro Clark thought that the situation in Macon County provided the opportunity for the study of untreated syphilis. On August 29, 1932, he sent a letter to Dr. J. N. Baker, State Health Officer in Montgomery Alabama, in which he wrote:

> In working up the data for the final report to the Julius Rosenwald Fund I was particularly impressed with the fact that a negligible number, something less than 35, of the Negroes under treatment in Macon County during the period of the demonstration had ever had any previous treatment. It seems to me that this situation in a very heavily infected population group affords an unparalleled opportunity of studying the effect of untreated syphilis on the human economy.

Baker's response came about a month later, on September 23, 1932. The Macon County Health Board, he wrote, had approved the project "... but with the distinct understanding that treatment will be provided for these people." Clark was perhaps indifferent to the health of those "Negroes," but it seems that the Macon County Health Board was not. They asked for treatment for those previously untreated people, and they had actually arranged that it be given by the people working in the Tuskegee institute.

Why did Clark propose the study of untreated syphilis? Historian James H. Jones has argued that Clark's leadership and hard work during the Rosenwald Fund's program leave no doubt that he would have preferred to return to Macon County to treat people rather than study those with untreated syphilis. However, with the Fund's withdrawal, there was no financial support for offering treatment. Therefore, the question that remained was whether there was anything else to be gained from this situation. And Clark thought that the study of untreated syphilis could offer new insights, particularly about the effects of the disease on Black people. This was something that had already been done with White people, with the results of such a study in Oslo by E. Bruusgaard having already been published in 1929. Interestingly, the Oslo study had arrived at the conclusion that neurological problems were rare in White people, which was exactly what was thought at the time about Black people. Therefore, the study of untreated syphilis in the Black people of Macon County would allow for a comparison between White and Black people, in order to figure out in a more systematic manner whether syphilis affected them in the same way.

However, there was a key difference between the Tuskegee and the Oslo studies. The latter was a retrospective study, that is, it was based on case histories rather than ongoing examinations. Clark believed that the Tuskegee study had a lot to offer in providing current evidence for the impact of syphilis on Black people. In this way, the study would not only show concern for them but would also provide evidence to support legislations in the South for funding treatment programs. Furthermore, Clark also knew that the study had risks for participants. Even though most physical examinations were harmless, in order to diagnose neural syphilis, it was necessary to perform lumbar punctures, by inserting a large needle into the spinal canal. This was a painful procedure that caused headaches, and in some cases paralysis or even death. Apparently, Clark concluded that these risks for participants were outweighed by the benefits of the study for science. As Jones noted, Clark's reference to the study as providing an unparalleled opportunity "… seemed to equate the absence of obstacles with a mandate. They were not the words of a man who entertained any ethical or moral qualms about what he was proposing."[6] In some sense, the fate of those people seemed fixed anyway, so whatever medical knowledge could be gained by the study would be an advantage. This was all done in the name of science, or so Clark believed.

Deceit … in the name of science

Historians Paul Lombardo and Gregory Dorr think otherwise. They point out that it is no coincidence that Clark himself, as well as Surgeon General Hugh Cumming who initiated Tuskegee, and Raymond Vonderlehr who directed it during its first decade, all had strong ties to racial medicine and eugenics. These were dominant movements in the beginning of the twentieth century, even though they started losing traction in the 1930s because the Nazi regime took those to the extreme. The key idea behind these movements was that racial groups were differentially susceptible to infectious disease, such as syphilis, due to inherent biological differences (racial medicine), as well as that society could advance if it promoted the breeding among "superior" people, while limiting the breeding among "inferior" people (eugenics). In the latter case, sterilization was also implemented as a measure in the USA and elsewhere (see Chapter 9). Lombardo and Dorr noted that Clark, Cumming, and Vonderlehr had all learned a brand of "racial medicine" that found "scientific" validation in eugenics, that this learning had occurred at the University of Virginia's medical school thus resulting in interpersonal relations that made the PHS a stronghold of Virginia physicians, and that during their PHS careers they were all publicly associated with the eugenics movement. For them, according to Lombardo and Dorr, the Tuskegee study provided the means for testing the eugenic ideal of existence of biological differences among races.[7]

Most crucially, Clark did not disclose the real objective of the study to the participating physicians and health personnel. Instead, he relied on the good impressions that the previous study under the auspices of the Rosenwald Fund had left, and implied that they would keep on working along the same lines—apparently not feeling the least embarrassed by the deceit. Thus, Eugene Dibble, the medical director of the Tuskegee Institute and head of Andrew Hospital, agreed to volunteer the services of his personnel to the study in order—as he thought—to administer syphilis treatments. This would be done under the supervision of Vonderlehr who was in charge of all the field work, being assisted by Eunice Rivers, a Black nurse who had been chosen by Dibble as a special scientific assistant for the study. Vonderlehr got to work, first aiming to identify those individuals with a positive test who had never received any treatment. Within a week, 300 people had been tested and found to have an infection rate of 17%, which was less than half of the 35% found during the Rosenwald Fund demonstration. In the beginning, Clark did not intend to leave participants untreated in the long term, and only had in mind a short-term study of a year or less. However, providing the necessary drugs and medical supplies was a burden that had to be dealt with by PHS. Vonderlehr, following the agreement with the local authorities, provided some minimal treatment, and at some point 500 people per week were receiving it. However, Clark was not pleased because treatment was not his main goal, especially as the one offered was not sufficient for curing people. Vonderlehr nevertheless insisted that a minimum treatment had to be maintained until the complete sample had been selected and examined. Ironically, the maximum treatment given was less than half the amount recommended by the PHS to cure syphilis!

Vonderlehr soon became excited by the findings of the physical examinations, without wondering much about the participants' well-being. When the time came for the spinal taps, he transported participants to a designated area in groups of 20 people, hoping that he would thus avoid having the others learn how painful this procedure was and thus decline to participate. Importantly, Clark did not object to the practice of deceit by not letting participants know about the procedure. The procedure had been presented to them as a special free treatment from the government, accompanied by free hot meals, free physical examinations, and free burial insurance. But the letter Vonderlehr sent to potential participants made no mention of lumbar punctures, or that the procedure they would undergo was diagnostic and not therapeutic—participants were instead told that they would be given "spinal shots"! Because many of them had received injections with neoarsphenamine as treatment, they also thought of spinal shots as therapy. By mid-1933, 307 men had undergone this painful procedure, which left a permanent mark of fear on many of them.

The experiment could have ended there, but Vonderlehr's succeeding Clark after the latter's retirement gave it a new boost. In a letter to Clark, written on April 8, 1933, Vonderlehr had written:

> At the end of this project we shall have a considerable number of cases presenting various complications of syphilis who have received only mercury and may still be considered untreated in the modern sense of therapy. Should these cases be followed over a period of from five to ten years many interesting facts could be learned regarding the course of complications [of] untreated syphilis.

Vonderlehr had already devoted nine months of hard work to the study and wanted it to continue before taking over from Clark. He was convinced of its merit for better understanding the impact of syphilis on the cardiovascular system of Black men, as well as that it could provide evidence that would support treatment programs for Black people. However, he overlooked its most important flaw: many of the participants had received some treatment. Vonderlehr did not, or did not want to, realize that the study might have some value as a study of the effect of undertreated syphilis; it was not a study of untreated syphilis. But the worst was yet to come.

On July 18, 1933, he wrote to O. C. Wenger, chief of the federally operated venereal disease clinic at Hot Springs, Arkansas:

I have also discussed the matter with a number of officers here in Washington and everyone is agreed that the proper procedure is the continuance of the observation of the Negro men used in the study with the idea of eventually bringing them to autopsy. I realize that this may be impracticable in connection with some of the younger cases, but those more advanced in age with serious complications of the vital organs should have to be followed for only a period of few years.

Wenger's response on July 21, 1933, is astonishing: "As I see it, we have not further interest in these patients *until they die*" (emphasis in the original). Performing autopsies on participants after they had died would provide additional important data for the study, as it would be possible to study directly the affected organs rather than indirectly infer their condition. It was thus crucial to have access to the bodies of participants as soon as they died. To achieve this, they needed to learn about the death immediately and access the body before it was buried. At the same time, the family would have to be convinced that major alterations to the body would not take place, even if they were, otherwise they would most likely not agree to the autopsy. So, deceit was used once again. It was sufficient that the patients died in the hospital, where the researchers would immediately learn about the death, and where they could perform the autopsy without any interference from the family. Here is how Wenger proposed this plan, in the same July 21 letter:

There is one danger in the latter plan and that is if the colored population becomes aware that accepting free hospital care means a post-mortem, every darkey will leave Macon county and it will hurt Dibble's hospital. This can be prevented, however, if the doctors of Macon County are brought into our confidence and requested to be very careful not to let the objective of the plan be known.

In essence, the idea was to let these people think that they would be offered medical care while all that they would get would be an autopsy after their death.

Vonderlehr shared the plan of the study with various experts. While he appreciated the positive responses, he entirely dismissed the questions and objections that he received. The American Heart Association totally rejected the scientific validity of the procedures and tests on which the diagnoses had been based. But Vonderlehr continued undistracted, considering this criticism as a simple difference in opinion. In an attempt to strengthen the study, he decided to add a control group so that appropriate comparisons could be made. Periodic examinations in both groups, without spinal punctures in the controls, were expected to allow for useful comparisons of the development of the disease. There was no discussion of ethical issues at any point. But there were methodological problems too. When 12 men in the control group were found in 1939 to be positive for syphilis, they were simply transferred to the syphilis group!!!

Astonishingly, nurse Rivers, who was otherwise very close to the study participants and cared for them, was found in the position to have to exempt them from treatment on at least two occasions: in 1937 when the Rosenwald Fund decided to renew their support of syphilis control programs and send a Black physician, Dr. William Perry, to Macon County; and in 1939 when a PHS mobile treatment unit visited the area. In both cases, Rivers indicated that the men in the study should not receive any treatment because they were under study. Unfortunately, Rivers took for granted that as a Black female nurse she was not to question the orders of the White male doctors. As she explained in an interview: "as a nurse being trained when I was being trained we were taught that we never diagnosed; we never prescribed; we *followed* the doctor's instructions!"[8]

When Dr. John Heller succeeded Vonderlehr as medical director of the Venereal Disease Division of the PHS in 1943, that was also the year that the PHS started administering penicillin as a treatment for syphilis to various clinics across the country. Therefore, the experiment could have been terminated then. But it was not. In a total irony, finding that penicillin could cure syphilis made the continuation of the Tuskegee study even more important, in order to provide a basis for comparison. Here is what Thomas Parran, Surgeon General of the United States from 1936 to 1948, wrote to Catherine Doran, Assistant Secretary of the Milbank Memorial Fund, on November 4, 1943, in a letter asking for the annual grant:

> This study, with its careful and complete physical examinations and subsequent observation up to and including autopsy at death, forms a necessary control against which to project not only the results obtained with the rapid schedules of therapy for syphilis but also the costs involved in finding and placing under treatment the infected individuals.[9]

For the PHS staff, the Tuskegee study participants were research subjects, not patients. In 1946, Heller reported what had happened until that time. Between January 1, 1932 and December 31, 1944, 101 out of 410 people of the syphilis group had died, compared to 28 out of 201 controls.[10]

Nobody offered any care to the people in the syphilis group until Peter Buxtun, once a PHS employee, shared the story with journalists who went public in 1972. It took another 25 years for the US government to officially apologize. Documentaries during the 1990s showed several of the federal doctors involved in the study who were still alive expressing little to no remorse. Angered by this, a group of historians and health providers, aided by the Black Congressional Caucus, successfully convinced President William J. (Bill) Clinton to offer a formal federal apology in 1997 to the last six surviving participants, their families, and the entire African American community.[11] Here is what Clinton had to say:

> To the survivors, to the wives and family members, the children and the grandchildren, I say what you know: No power on Earth can give you back the lives lost, the pain suffered, the years of internal torment and anguish. What was done cannot be undone. But we can end the silence. We can stop turning our heads away. We can look at you in the eye and finally say on behalf of the American people, what the United States government did was shameful, and I am sorry.
>
> The American people are sorry—for the loss, for the years of hurt. You did nothing wrong, but you were grievously wronged. I apologize and I am sorry that this apology has been so long in coming.
>
> To Macon County, to Tuskegee, to the doctors who have been wrongly associated with the events there, you have our apology, as well. To our African American citizens, I am sorry that your federal government orchestrated a study so clearly racist. That can never be allowed to happen again. It is against everything our country stands for and what we must stand against is what it was.[12]

As James Jones and Susan Reverby, two historians who have studied this story very closely, commented: "That was an important gesture, but it could not end the mistrust. It was only one step toward truth and reconciliation. It merely acknowledged the past. It did not change the present nor shape the future."[13]

One of the positive impacts of the Tuskegee study was that it made the US government change its research practices. In 1974, the National Research Act was signed into law, creating the National Commission for the Protection of Human Subjects of Biomedical and Behavioral Research, which defined basic principles of research conduct and ensured that they were followed. In the same year, regulations were passed that required researchers conducting studies funded by the Department of Health, Education, and Welfare to get voluntary informed consent from all persons taking part in studies and have all the research protocols reviewed by Institutional Review Boards that would decide whether they met ethical standards. These were just the first of several measures taken since.[14] However, there was also a very negative impact that the Tuskegee study has had on public perceptions of medical science and participation in related research, particularly but not only among African Americans. This has been documented in various studies that are too many to consider here. Worse than that, the Tuskegee study is not the only case of discrimination and exploitation of African Americans. Historian Harriet Washington has described the whole situation as "Medical Apartheid" (the Apartheid was the racial segregation under the all-White government of South Africa, which lasted from 1948 to 1994 and which dictated that the majority of non-White South Africans were required to live separately from Whites). As she put it, "To gain trust, we first acknowledge the flagrant abuses of the past and the subtler ones of the present ...."[15] But mistrust among people of African origin remains, independently of Tuskegee. The situation has improved significantly in recent decades, but problems still remain. During the COVID-19 pandemic, Black people in the USA were quite reluctant to get vaccinated. As it was nicely put, "... the best way to learn from the atrocities of the past is to change our present."[16]

Given the Tuskegee study story and other cases of medical discrimination and exploitation of African Americans, it should not be surprising to hear about conspiracy theories involving scientists who act against particular groups. Indeed, it has been found that one in two people in the USA accept at least one medical conspiracy theory. In a study, a bit more than one in three agreed that the Food and Drug Administration is intentionally suppressing natural cures for cancer because of drug company pressure, whereas one in five agreed either that corporations were preventing public health officials from releasing data linking cell phones to cancer or that physicians still want to vaccinate children even though they know such vaccines to be dangerous.[17] This may sound exaggerated, but if you look from the side of people from oppressed and exploited groups, the question does make sense: If the government could maintain the Tuskegee study for 40 whole years with most people not knowing about it, and if it eventually became known only thanks to a whistle-blower, why doubt that they are capable of successfully implementing other medical conspiracies?

Before we can answer this question or decide whether what the PHS people did to the Tuskegee people was indeed a conspiracy, it is necessary to clarify what a conspiracy is.

Conspiracies and conspiracy theories

In simple terms, we can say that a conspiracy exists when a *small group* (the conspirators) operates *in secret* to do something *harmful*. One could argue that the Tuskegee study had all these features: Some people in the PHS (hence a small group) convinced the local authorities in Alabama without disclosing all the facts (hence the secrecy) to keep the Tuskegee study participants untreated for syphilis even after an effective treatment was made available (hence doing harm). A well-known conspiracy concerns the events of September 11, 2001. On that day, 19 terrorists hijacked four commercial airplanes and crashed two of them into the Twin Towers of the World Trade Center in New York City, another one onto the Pentagon, the headquarters of the U.S. Department of Defense in Arlington, Virginia, whereas the fourth plane went down in rural Pennsylvania. The outcome was the death of about 3,000 people. The official account of the 9/11 events and of how this small group of 19 terrorists worked secretly to prepare the attacks and do harm is a tale about a real conspiracy. Conspiracies do happen, and a conspiracy theory is simply a tale about a conspiracy, which might or might not be true. But there also exist what philosopher Quassim Cassam has helpfully designated as Conspiracy Theories (CTs): theories that are entirely implausible and that serve to advance an ideological position.[18]

A popular CT has been that HIV, the virus that causes AIDS, was a bioweapon; this was a story that was the outcome of a massive disinformation campaign by the Soviet KGB. An alternative version is that HIV is harmless and that the drugs offered against it are the real cause of AIDS. The history of racialized medical abuse in the United States, including the Tuskegee study, and elsewhere has provided a fertile ground for AIDS CTs. For instance, in South Africa, President Mbeki and Minister of Health Tshabalala-Msimang promoted AIDS conspiracy stories while at the same time delayed offering antiretroviral drugs to those people who needed them. Mbeki supported dissident scientists (see Chapter 6), alleged that the CIA and the large pharmaceutical companies were part of a conspiracy promoting the view that HIV was the only cause of AIDS and that toxic antiretroviral drugs were the only treatment, and along with Tshabalala-Msimang supported the replacement of these drugs with alternative cures and nutritional interventions, with devastating consequences.[19]

Several conspiracy theories also became popular during the COVID-19 pandemic. A topic that figured prominently in many of these was the origin of the SARS-CoV-2 virus: Did it come from animals (zoonotic origin hypothesis) or was it due to a laboratory-related accidental leak (lab leak or human origin hypothesis)? The latter hypothesis is not entirely unreasonable given how close the Huanan wet market in Wuhan, where many of the earliest cases of COVID-19 were identified, is to the Wuhan Institute of Virology that conducts research on coronaviruses. Four years since the beginning of the pandemic, there are scientists who argue on social media that a massive cover-up took place in February 2020. For instance, Bryce Nickels, Professor of Genetics at Rutgers University wrote on his X (formerly Twitter) account on February 1, 2024: "Whether SARS-CoV-2 came from a lab or from nature, what is indisputable is that during a conference call four years ago today (February 1, 2020) a decision was made by unelected officials & scientists to intentionally deceive the public about the origin of the COVID-19 pandemic."[20] Sharing Nickels' post, Jay Bhattacharya, Professor of Health Policy at Stanford University, wrote: "This is the biggest scientific scandal of all time. In Feb. 2020, top science bureaucrats in the US and the UK conspired to cover up the legitimate possibility that covid was caused by a pandemic preparedness industry lab leak."[21] Could this be true? Could there have been a massive cover-up about the real origin of SARS-CoV-2? And if yes, why did it happen?

In reviewing the available evidence, Stephan Lewandowsky, Peter Jacobs, and Stuart Neil have concluded that although the question about the origin of SARS-CoV-2 is not entirely settled, the evidence for a lab leak up to that point was not convincing. According to them, most virologists have supported a zoonotic origin of SARS-CoV-2, with the most convincing evidence coming from the comparison of the genomes of SARS-CoV-2 and other similar viruses. The comparisons clearly show how SARS-CoV-2 could have evolved from other viruses, and this evidence was made publicly available early on during the pandemic.[22] A detailed analysis of the locations of the very first infections in Wuhan in December 2019 has provided additional evidence for a zoonotic origin of SARS-CoV-2. That less than two-thirds of the early cases were linked to the Huanan market is not surprising, given how highly transmissible SARS-CoV-2 is and how high the rate of asymptomatic transmission is (that is, how many people are infected without showing any disease symptoms). However, it is no coincidence that many of the more than 100 COVID-19 cases from December 2019 with no epidemiologic link to the Huanan market nevertheless lived very close to it. This is compelling evidence that the Huanan market was where the SARS-CoV-2 transmission began.[23] This evidence in turn weakens the arguments for a lab leak. There also exist conspiratorial views that SARS-CoV-2 was designed as a bioweapon that was released in Wuhan. Lewandowsky, Jacobs, and Neil argue that the Huanan market "natural origin" and the "lab leak" hypotheses crucially depend on the evidence for or against the evolutionary origin of SARS-CoV-2. The more the evidence for a "natural" evolutionary origin, the weaker is the support for a lab leak—which however can never be entirely ruled out in principle.[24]

This was further supported by a geographically diverse, and anonymous (very important to note!), survey of scientific experts regarding the origin of the COVID-19 pandemic, which has shown that most of them agree with the view of a natural origin of COVID-19. The study involved 168 virologists and infectious disease epidemiologists, as well as some biosafety/biosecurity professionals and evolutionary geneticists, drawn from 47 countries all over the world. When they were asked how likely it is that COVID-19 originated from "natural zoonosis" (an infectious agent causing disease that is transmitted from an animal to humans), four out of five of these experts stated that such an origin was more than 50% likely, with the rest reporting a 50% or greater chance for another kind of origin. In other words, even though there was no complete consensus among these experts, most of them believed that a natural origin was more likely than a lab-related accident. Interestingly, about nine out of ten experts also believed that more research on this topic is required.[25] As already mentioned, the lab leak hypothesis for the origin of SARS-CoV-2 is not entirely unreasonable, but the currently available evidence does not support it as the most plausible explanation.

This notwithstanding, when scientists in prominent universities, like Nickels and Bhattacharya, insist that there has been a conspiracy, it is reasonable for lay people to be concerned. So, what they need is to be able to distinguish between plausible, albeit less likely, conspiracy theories, such as the lab leak origin of SARS-CoV-2, and CTs that are entirely implausible and unlikely, such as the theory that HIV does not cause AIDS whereas the drugs administered against it do. But do lay people believe in CTs? Indeed, they do! For instance, a 2020 study in the USA found that about one in three people agreed that the virus was purposefully created and spread, as well as that the threat of COVID-19 had been exaggerated to damage President Trump. Interestingly, with respect to the "spread on purpose" theory, those who exhibited the lowest degrees of conspiracy thinking usually disagreed with this view, whereas those who showed the highest level of conspiracy thinking on average agreed.[26] In another study in 17 European countries, it was found that one in five participants who held conspiracy beliefs had lower COVID-19 vaccine uptake, were less satisfied with the way health services dealt with the pandemic, and were less supportive of governmental restrictions. One of the items used to measure conspiracy beliefs was "Coronavirus is the result of deliberate and concealed efforts of some government or organization," with about one in four participants agreeing or strongly agreeing with this statement.[27] A study of six countries in Asia and Africa found that about one in three participants believed that SARS-CoV-2 was a synthetic virus, about one in five believed that COVID-19 vaccines were a way of implanting people with microchips and that the vaccines would lead to infertility, about one in four thought that immunizing themselves and their children was harmful and that this fact was covered up, and about half were convinced that pharmaceutical companies were not honest and hid the dangers of COVID-19 vaccines, thus being seriously concerned about their safety and efficacy data.[28] Finally, a study of people in 13 Southern American countries found that between one in four and one in two people agreed with the CTs about vaccination against COVID-19.[29] In short, CTs are accepted by many people all over the world.

My assumption is that CTs like the aforementioned ones often reflect genuine and legitimate concerns that lay people have. But because their factual claims are not solid and because they can do real harm, for instance, by making people refrain from getting vaccinated, it is necessary for lay people to be able to recognize and identify them. The question then becomes: How can this be done?

What is crucial to understand? (and also teach in schools!)

In order to be able to identify CTs, we need a set of criteria that can help us distinguish between them and other, plausible in principle, conspiracy theories. The Tuskegee study has shown that conspiracies involving scientists are possible, even though this story was more the outcome of prejudice, indifference, and negligence rather than of a real conspiracy. Whether there is a conspiracy to cover up the human causes of COVID-19 remains to be seen; it currently does not seem likely, but it is not entirely impossible either. But real conspiracies do exist. What happened in the USA on September 11, 2001 was the outcome of a conspiracy among a small group of conspirators who flew planes onto the Twin Towers of the World Trade Center in New York, and onto the Pentagon in Washington, DC (with another plane going down on the same day). This does not entail that all CTs are true, but only that conspiracies do happen.

Psychologist Jan-Willem van Prooijen has suggested that in order for a belief to qualify as a conspiracy theory, it has to have at least the following five critical features:

1. *Patterns*: Events are explained by assuming nonrandom connections among actions and events, rejecting the idea that they could be just coincidences.
2. *Agency*: The rejection of coincidences is justified because the events explained are the outcomes of the intentional and purposeful action of conscious agents.

3. *Coalitions*: These agents act in coordination and collaboration; there is no conspiracy if only one individual is involved.
4. *Hostility*: These agents also act in order to achieve goals that are evil, selfish, unlawful, or otherwise not in the public interest.
5. *Continued secrecy*: These coalitions of malevolent agents act in secrecy, and this is why they remain in the realm of theory. When a conspiracy is shown to occur, it is no longer a theory but a fact.[30]

A classic historical example of a real conspiracy was the "Wannsee conference." On January 20, 1942, 15 high-ranking Nazi Party and German government officials, including SS General Reinhard Heydrich, the chief of the Reich Security Main Office and one of the top deputies to the SS Chief Heinrich Himmler, as well as SS Lieutenant Colonel Adolf Eichmann, chief of the Department of Jewish Affairs, gathered at a villa in the Berlin suburb of Wannsee. The goal was to discuss and coordinate the implementation of what they called the "Final Solution of the Jewish Question." The decision for the systematic, deliberate, physical annihilation of the European Jews had already been taken before the Wannsee meeting. Its goal therefore was not to decide whether the Final Solution would be implemented, but how. Heydrich wanted to inform participants that Hitler himself had tasked him with coordinating the operation, and to secure the support of government ministries and other relevant agencies. However, the actual order had not been given to Heydrich and it depended on getting approval of his plan. This is why the Wannsee meeting took place. According to the surviving minutes of the meeting, the agenda included the precise definition of exactly which group of people was to be affected, followed by a discussion of how these people were to be deported and subjected to the toughest form of forced labor, as well as about how the survivors of this forced labor would be dealt with. The mass deportations had been initially scheduled to occur after the war, but it was decided that they should eventually occur during it. The details of the implementation of the final solution were not finalized during the Wannsee meeting, but in the following months.[31] In short, a small group (secrecy) of officials (coalition) convened to decide how to implement their plan (agency) of the systematic extermination (hostility) of particular groups of people throughout the occupied territories (patterns).

Conspiracies do happen, as already mentioned. Therefore, it is important to be able to distinguish between real or plausible conspiracies on the one hand and CTs on the other. To achieve this, it is helpful to look for particular features of the latter. Political scientist Michael Barkun has argued that CTs usually begin from three fundamental assumptions:

- *Nothing is as it seems*: Conspirators wish to deceive in order to disguise their identities or their activities, so all appearances are deceptive. Thus, the appearance of innocence or noninvolvement is no guarantee that an individual or a group is benevolent or did no harm.
- *Nothing happens by accident*: Conspiracies imply a world based on intentions, in which there is no place for accidents and coincidences. Anything that happens occurs because it has been intended by the conspirators. This is a world governed by design rather than by chance.
- *Everything is connected*: Because there is no place for accidents or coincidences, a pattern is believed to be everywhere, but is hidden. Therefore, one needs to engage in a continuous process of finding correlations in order to figure out the hidden connections.[32]

CTs are also of two kinds, according to Barkun. On the one hand, there exist "event conspiracies," that is, conspiracies that have brought about a single event or a set of events. On the other hand, there exist "systemic conspiracies," that is, conspiracies that have broad goals and aim at controlling a country, a region, or the whole world.[33]

Perhaps the most well-known event CT is that Lee Harvey Oswald did not kill President Kennedy alone. According to these CTs, this was not the case of a lone wolf making history. There is no way that Oswald could have set the assassination up on his own and been successful (*Nothing is as it seems*). Furthermore, there is no way that Oswald was lucky enough to have the right angle and the necessary time in order to shoot Kennedy three times within a few seconds. Rather, he must have had other accomplices who either set things up for him or even shot Kennedy too (*Nothing happens by accident*). Finally, it was not a coincidence that the assassination happened when it happened. Kennedy had many enemies at the time (the Soviets, the Cubans, the mafia, and the list goes on) who might have conspired to take him out of the way (*Everything is connected*). For many people, it makes more sense to explain the Kennedy assassination as the outcome of a well-organized conspiracy of malicious people intending to take out a charismatic leader than of the actions of a single individual who just happened to be able to successfully implement his plan. Subsequent events, such as Oswald's assassination a couple of days later have fueled CT thinking. Many people believe that Oswald was killed in order to avoid a trial and thus keep the underlying conspiracy a secret. However, for those who have looked closely at the evidence, there is no evidence that there was no conspiracy behind Kennedy's assassination. Therefore, it is closer to being a CT rather than a plausible conspiracy theory.[34]

A well-known systemic CT is the one about the *Protocols of the Elders of Zion*, the minutes of a meeting that supposedly set the agenda for a Jewish plan for global domination. The key idea was that an organized group of Jewish Elders conspired to bring about a Jewish dictatorship by undermining the current world order; that this was achieved through the proliferation of ideologies such as republicanism, liberalism, socialism, and anarchism; that these organized Jews controlled the economy and the press; that they were working in the background to undermine the beliefs of non-Jewish people and control them; and that major clashes between nations were the outcome of the machinations of the Elders of Zion. Even though it was shown very early that this document was a hoax, being a product of forgery and fabrication in Russia in the early twentieth century, it has since been translated in several languages and is still considered as a legitimate document by some anti-Semitic groups. Interestingly, its problems were well known to the Nazis who were reluctant to refer to this document in their own anti-Semitic propaganda.[35]

And there is more. Philosopher Quassim Cassam has identified some additional features of CTs that should be extremely helpful in identifying them. CTs are:

1. *Speculative*: They are based on conjecture, rather than solid evidence.
2. *Contrarian*: They are contrary to the official view of things, which they see as a way to cover up the conspiracy the CTs are intended to expose, as well as to any obvious explanations of events.
3. *Esoteric*: They appeal to the less obvious and to the more unlikely, putting no limits to imagination.
4. *Amateurish*: Their proponents are not experts in the related domains, and they do not have the qualifications required to express a legitimate opinion.
5. *Inherently purposeful*: Things always happen for a reason, and coincidences are always unacceptable as explanations.
6. *Self-sealing*: The arguments that support them and the grounds for their plausibility are not easy for outsiders to criticize, as people with differing assumptions can easily end up talking past one another.[36]

If you come across theories having several of these features, then you have good reasons to disregard them as CTs.

But this is neither straightforward nor easy to do, because quite often we find CTs very intuitive. As van Prooijen explains, CTs appeal to normal psychological processes, the perception of patterns, and the detection of agency. Negative emotions further amplify these feelings, and CTs thus become difficult to resist.[37] So, what can we do about this? Historian Michel

Jacques Gagné, who has carefully and diligently analyzed the John F. Kennedy assassination CTs that he accepted himself in the past and that he managed to overcome, has suggested the following simple steps:

1. Be ready to follow logic and evidence wherever these lead, even at the risk of being wrong.
2. Respectfully engage with people holding different views, especially when these are experts in their domain, and do not think of official investigations as a priori attempts to cover up something.
3. Refrain from speculating too much about what else could have happened besides what logic and evidence point to.[38]

Conspiracies do occur, but they are most likely to be creations of our mind; therefore, we should thoughtfully and carefully consider the evidence for them. Next time you come across a conspiracy theory related to science, you can try to apply the criteria just described in order to decide for yourself whether it is worth considering or not.

4

"The Compulsory Vaccination Act ... invades ... the liberty of the subject, and the sanctity of home"

Individual liberties and bodily autonomy

The "Minister of Death"

> Sir, I humbly beg to protest against the passing of the Bill on Compulsory Vaccination, which I look upon as a gross infringement on the Medical liberty of the Subject. I further beg to represent to you, that many professional persons and others are of opinion that Vaccination not only does not prevent small-pox, but it is productive of worse and more dangerous diseases, such being the case, I hope you will present this my protest to the House of Commons.[1]

This is a letter from May 5, 1856, sent by Andrew Lowden to Sir George Grey, Secretary of State in Great Britain, protesting against the compulsory vaccination bill.[2] The two main arguments in this letter, about individual liberties and vaccine effectiveness and safety, have characterized most public anti-vaccination movements since then. Vaccine effectiveness and safety will be considered in subsequent chapters of the present book. In this chapter, I focus on the question of individual liberties. This is a key cause of distrust in science even though it is only indirectly related to science per se.

The problem that the compulsory vaccination bill aimed to address was smallpox, a disease that had killed millions of people over thousands of years. The word "pocks," later changed to "pox," was used to describe any unpleasant eruption on the skin. The term "small pocks" (later small-pox, and finally smallpox) was coined in order to distinguish this disease from the "great pox" or syphilis. Smallpox is caused by the variola virus (VARV), a DNA virus (see Box 2.1). The symptoms of the disease it caused were really bad: high fever, vomiting, and mouth sores, followed by fluid-filled lesions on the whole body.

Trusting Science. Kostas Kampourakis, Oxford University Press. © Oxford University Press (2025).
DOI: 10.1093/oso/9780197787106.003.0004

About half of those infected died, some of them within two weeks. Those who did not die could be left with permanent problems such as disfigurement and blindness. Besides the main variety of smallpox viruses, called *Variola major*, there is another less lethal one called *Variola minor*. Those infected by *V. minor* were lucky both because the disease they developed was mild and because they were thereafter protected against *V. major*.

The first method used to protect people from smallpox was variolation, that is, the inoculation of uninfected people with human smallpox. Variolation was introduced to Great Britain in the 1720s by Lady Mary Wortley Montagu, wife of the British ambassador to the Ottoman Empire. Mary had lost her young brother to the disease in 1713 and had suffered from it herself in 1715—an incident that left her with many pockmarks on her face, whereas she had been previously noted for her beauty. She therefore had her son Edward inoculated in 1717 in Istanbul and her daughter Anne in 1721 after their return to London. In fact, Anne was the first person to be inoculated in England. What was done in variolation was either to insert smallpox under the skin or have people inhale smallpox scabs. These practices usually resulted in a mild form of the disease and subsequently provided protection from smallpox. However, in some cases, people developed a severe form of the disease and died. So, a safer method was needed, and it was a physician from Gloucestershire, Edward Jenner, who popularized an alternative one.

In May 1796, a farmer's daughter named Sarah Nelmes consulted Jenner because of a rash on her hand. Jenner's diagnosis was that it was cowpox. When Sarah confirmed that one of her cows had recently had the disease, Jenner realized that infection from cowpox, which was a mild disease for humans, might confer protection against smallpox. He thus decided to test this idea by giving cowpox to someone who had not yet suffered from smallpox. Jenner extracted pus from Sarah's hand and applied it through small incisions to the skin of eight-year-old James Phipps on May 14, 1796. Even though he reacted to the cowpox matter and felt unwell for several days, James made a full recovery. On July 1, 1796, Jenner inoculated the young boy with human smallpox to test his resistance, and indeed James remained in perfect health. It should be mentioned that Jenner himself had been inoculated with smallpox when he was younger and got through only a mild version of the disease (which might be a reason why he did not do the experiment on himself, as he might have thought that he was already protected).

Jenner tried to have his conclusions published in a paper in the *Transactions of the Royal Society*, which was rejected as it was considered to be based on insufficient evidence. His conclusions were thus made public in a short 1798 book titled *An Inquiry into the Causes and Effects of the Variolae*

Vaccinae. He presented 23 cases in that book, including Sarah Nelmes (case 16). He considered his observations as providing sufficient evidence for the conclusion that "... the Cow-pox protects the human constitution from the infection of the Small-pox."[3] It is important to note that it is not entirely clear whether Jenner inoculated James with cowpox, inactivated smallpox, or something else between them. Most interestingly, in his 1798 book, Jenner referred to a cowpox virus; however, he used the term "virus" in the sense that it had since older times of some harmful factor rather than in the modern sense of an acellular pathogen. The English word "virus" comes from the Latin word "vīrus," which refers to poison, and was initially used to refer to an unknown cause of disease other than bacteria. Jenner did not know, and could not have known, that smallpox is caused by the VARV.

Even though Jenner was not the first to suggest that cowpox could protect against smallpox, he was the first to provide some empirical evidence about this, outrageous and disturbing from an ethical point of view as Jenner's experimentation with a young boy was. It is worth mentioning that the royal family had done a similar experiment already in the 1720s. During the smallpox epidemic of May 1721, the daughter of the Prince and Princess of Wales fell ill. Perhaps because she thought it was smallpox, the Princess of Wales, Caroline of Ansbach, might have become interested in the new procedure of inoculation. Caroline knew Mary Montagu, but it is unclear what the latter's role and influence was in the former's decision to have two of her daughters inoculated. Within a month of the illness of the royal princess, the newspapers mentioned that some physicians had asked the King for permission to conduct smallpox inoculation experiments on condemned criminals in Newgate Prison, under the condition that they would subsequently receive pardon. The King apparently agreed, and thus three male and three female prisoners were inoculated on August 9, 1721. The experiment was conducted by surgeon Charles Maitland, who had also inoculated Montagu's children, was witnessed by at least 25 physicians and surgeons, and received extensive coverage by the press. The procedure was repeated three days later, and after a brief illness all patients recovered. These results convinced most people that variolation was rather safe, even though questions remained about what kind of protection it conferred. Eventually, the two princesses were successfully inoculated in April 1722.

Despite Jenner's success, not everyone was thrilled by the idea of vaccination, as it was deemed unnatural because it involved inserting material from animals into humans. Opponents of vaccination included physicians, such as Benjamin Moseley who already in 1800 had argued against this practice, asking his readers: "Can any person say what may be the consequences of introducing the *Lues Bovilla*, a bestial humour-into the human frame, after a long lapse of years? Who knows, besides, what ideas may rife, in the course of time, from a brutal fever having excited its incongruous impressions on the brain?"[4] It seems that this debate gave James Gillray, one of the first professional caricaturists in England, the opportunity to represent the anxiety inspired by the vaccination debates in a caricature he produced in 1802 (Figure 4.1). It shows Edward Jenner inoculating patients in the Smallpox and Inoculation Hospital at St. Pancras. Cow heads are shown to grow from parts of patients' bodies following vaccination. Patients are spoon-fed "opening mixture" as they come through the door. A boy standing next to Jenner is holding a pot with the label "vaccine pock hot from ye cow"; on his jacket there is a badge saying "Pancras"; and in his pocket is a paper entitled "Benefits of the vaccine process."[5]

It was soon observed that there were no smallpox outbreaks in communities with high rates of vaccination. In London, the number of cases of smallpox had dropped significantly by 1820, while the number of cases for other diseases such as measles was rising. This was evidence that vaccination

The Cow-Pock _ or _ the Wonderful Effects of the New Inoculation! _ Vide the Publications of ye Anti Vaccine Society.

Figure 4.1 A 1802 caricature showing Edward Jenner vaccinating patients in the Smallpox and Inoculation Hospital at St. Pancras, with the patients developing features of cows. The caption of the caricature reads "The cow-Pock—or—the Wonderful Effects of the New Inoculation!—vide, the publications of ye Anti-Vaccine Society." Colored etching, 1803, after J. Gillray, 1802. Wellcome Collection. Public Domain Mark.

for smallpox, and not sanitary measures, was making a difference. Otherwise, the cases of measles, for which there was no vaccine available, would have also dropped if sanitary measures had been the cause for the drop in the cases of smallpox.[6] Nevertheless, a smallpox epidemic between 1837 and 1840 resulted in thousands of deaths in England. To deal with this problem, in 1840, the government started offering the vaccine for free, while special attention was given to reaching poor people. The 1840 Vaccination Act also outlawed variolation due to concerns that it risked spreading smallpox rather than helping control it.

But the measure of free vaccination did not work as well as it had been hoped. In 1853, epidemiologist Edward Seaton, Secretary of the "Committee on Smallpox and Vaccination" of the Epidemiological Society of London, sent a letter to Home Secretary Viscount Palmerston, along with a copy of a report on the state of smallpox and vaccination in England, Wales, and other countries. Seaton estimated that the annual average number of cases of smallpox in the United Kingdom would not fall short of 100,000, adding that "Such is the humiliating result of our own apathy 50 years after the discovery of vaccination." For him "… the great obstacle to the universal diffusion of vaccination …" was the neglect combined with ignorance and prejudice, which was found "… only among the lower and uneducated classes of the community." These people usually objected to vaccination, Seaton continued, supposedly on the grounds that it was not "a sufficient safeguard," or that it was thought of as "… the means of introducing other diseases into the system." However, not only was there no strong foundation for these views, but he also doubted that this was what those people really thought. Rather, he considered this an excuse for "… their own indolence and indifference." These were the real obstacles, Seaton added, and the proof of this was how those same people rushed to get vaccinated when new cases of smallpox appeared where they lived, showing that there was "… no rooted objection to vaccination in the minds of the people."[7] For Seaton, the situation was clear. On the one hand, smallpox was an extremely contagious and dangerous disease, often a fatal one. On the other hand, vaccination had been shown for decades since Jenner to be an effective prevention against smallpox. Therefore, compulsory vaccination was for Seaton and the Epidemiological Society the best, if not the only, choice.

Compulsory vaccination as coercion

Despite the availability of vaccination, there were still many cases of smallpox. But why did people not have their kids vaccinated against a disease that had already killed millions? One reason was that parents were reluctant to have lymph from animals transferred to the bodies of their children, as it would taint the perceived "purity of the blood." Furthermore, there were concerns that calf lymph might also be the means for the transmission of animal diseases. These concerns were related to anxieties about human nature and our relation to animals. But even if it were human, not animal, lymph that was transferred, there were still concerns. For those parents who could not choose the vaccinifer, that is, the person from whom lymph would be taken, there was concern about its quality. Whereas some private practitioners used calf lymph supplied by the National Vaccine Establishment, public vaccinators were urged by the government to use material from one child to vaccinate others in order to maintain the supply of vaccine material. But this method also allowed for blood-borne diseases to be transferred during the process. There was no way to be certain that the lymph transferred did not also contain blood. Finally, the method used for vaccination at the time was painful. Vaccinators used a lancet to make cuts on the skin, at least at four different places on the arm, in which they then transferred the vaccine matter. These cuts could leave permanent scars on the previously unblemished body of an infant, something that made parents distressed. In short, vaccination affected two aspects of health that Victorians considered crucial: the purity of the blood and the integrity of the body. These concerns were important and not unfounded.

The authorities however dismissed these concerns. What mattered for them was the growing number of smallpox cases. Something had to be done about it and the government decided that it would have to be a vaccine mandate. The Vaccination Act of 1853 made vaccination compulsory (and free) in England and Wales for all children during the first three months of their life. This law targeted the lower classes mostly, because they were considered to have not-so-good hygiene and because they usually could not afford the necessary medical care. The main goal was to ensure the vaccination of those children whose parents were not profiting from the already freely available vaccination. The Registrar of Births and Deaths in every district would notify parents that they should have their newborns vaccinated. The parents would either return a vaccination certificate signed by a medical professional, or they would be fined a maximum of 20 shillings (1 pound, which corresponds to about 105 pounds today), which could be more than a week's salary for some people.[8] Whereas this measure was relatively effective in increasing the vaccination rates and in decreasing the number of smallpox cases, the results were modest as the law was not really enforced.

Worse than that, compulsory vaccination initiated widespread reaction. For many people, this was an attack on their liberty, as individuals no longer had the right to make their own decisions about their own bodies. The first reaction came from the hydropath John Gibbs in an 1854 pamphlet titled *Our Medical Liberties*:

> The Compulsory Vaccination Act, while dishonouring science, invades in the most odious, tyrannical, and, speaking as a Briton, unexampled manner the liberty of the subject, and the sanctity of home; unspeakably degrades the free-born Briton not only in depriving him of liberty of choice in a personal matter, but even in denying him the possession of reason; outrages some of the finest feelings and best affections of the human heart;—those feelings and affections which have their origin in parental love—that still bright spark of the Divine Nature breathed into man by his Heavenly Father;—sets at nought parental authority and responsibility, and coerces the parent either to violate his deliberate, cherished, and conscientious convictions, and even his religious scruples, or boldly to defy an unjust and tyrannous law.[9]

Gibbs effectively summarized in one paragraph all the main issues: personal liberty was attacked, and home life was invaded. Parents were thus put in a difficult position: "damned if you do and damned if you don't" as the saying goes, because they either had to accept coercion and go against their own convictions or defy the law and deal with the consequences.

In the 1860s, the number of smallpox cases continued to rise, and so a new Vaccination Act in 1867 made vaccination against smallpox compulsory for all children under the age of 14. The Act indicated that any child under 14 years old who was found to still be unvaccinated should be vaccinated immediately. In case this was not done, there was again a fine of 20 shillings maximum to be paid. In addition, anyone who attempted variolation for whatever reason would be imprisoned for a period of one month maximum.[10] What was new in this Act was not only the age range of children that ought to be vaccinated, which was expanded from the age of 3 months to the age of 14 years (apparently to cover all children born since 1853 who were still unvaccinated 14 years later in 1867). It was also that the administrative machinery for policing and enforcing the law was by that time in place. A new clause was added that precluded parents from refusing vaccination. Each order to have one's child vaccinated that was disobeyed constituted a new offense and could therefore be the subject of a new conviction and penalty.

This Act brought about new reactions. The first anti-vaccination society to be founded was the Anti-Compulsory Vaccination League (ACVL), which was founded by Richard Butler Gibbs, a cousin of John Gibbs. It is important to note that among the ACVL members, there were physicians repeating the two key arguments against vaccination that we have considered already. For instance, Dr. Caplin noted that he objected to compulsory vaccination because in his own medical practice he had patients whose blood had been contaminated during the vaccination procedure. Dr. Epps of the Royal Jennerian and London Vaccine Institution who had vaccinated more than 120,000 children, and who was aware of the protection that vaccination conferred against smallpox, argued that nobody had the moral right to enforce vaccination upon people who objected to it. He considered vaccination to be the introduction of a poison into the system for the purpose of overcoming another poison, on the principle of choosing the lesser of two evils.[11] By 1870, the ACVL had dozens of branches and thousands of members. However, the death of Richard Butler Gibbs in 1871 negatively affected the movement.

The Vaccination Act of 1871 aimed to address concerns about the reluctance to take legal action against parents who refused vaccination for their children. This act established the appointment of vaccination officers with the specific task of ensuring compliance within their areas. Vaccination officers would also take over the key paperwork, such as keeping the vaccination register updated and issuing the certificates. However, this approach leaned too heavily towards law enforcement. Even supporters of compulsory vaccination were troubled by cases of working fathers who were imprisoned for a short period because they were circumstantially unable to vaccinate their children promptly. Such harsh penalties garnered public sympathy for those opposed to compulsory vaccination. This became especially evident in cases of distraint sales; when a person could not afford to pay the fine for noncompliance with the Vaccination Act, their goods could be seized and sold at auction. This generally happened only to the working class because these were usually the people who could not afford to pay the fines. Distraint sales also offered an opportunity for protest to anti-vaccination groups, who mobilized their members and supporters and organized demonstrations at the auctions, often resulting in violent clashes.

The year 1871 was also when the number of fatalities from smallpox rose significantly. There were 9,837 deaths by smallpox per 100,000 deaths from all causes in London, compared to 1,253 in 1870.[12] This was hard to explain. On the one hand, Seaton attributed deaths to the neglect of vaccination, or its inefficient application; those who were not vaccinated died. On the other hand, anti-vaccinationists argued that this high number of deaths showed that the advantage that vaccination supposedly conferred was an illusion; there were still many deaths due to smallpox, despite the enforcement of vaccination.

Even for those parents who accepted compulsory vaccination, there were still concerns about the safety of the procedure. Public vaccinations were generally performed at vaccination stations (Figure 4.2), which many parents considered as sites of pollution. Vaccinators performed numerous vaccinations per day, often from child to child: Material from a previously vaccinated child was used to vaccinate another one. But they were not really concerned about fully inspecting the children for signs of diseases, such as syphilis, which could be transferred from one to another. Thus, for those parents who could not afford a private practitioner, there was the fear of their child getting a disease by receiving vaccine matter from a child of unknown condition. Vaccinators were paid per child and had many children to vaccinate; as a result, they had little time for parents' legitimate concerns.

THE DISTRICT VACCINATOR—A SKETCH AT THE EAST-END

Figure 4.2 A dispensary in the East End of London: Crowds of local children are being vaccinated. Wood engraving by E. Buckman, 1871. Wellcome Collection. Public Domain Mark.

In 1874, the National Anti-Compulsory Vaccination League (NACVL) was formed by Mary Hume-Rothery, a writer and campaigner for medical reform. However, their approach to things alienated many people. For instance, whereas the NACVL asked working-class people to fund their association, they denied them any meaningful role in the anti-vaccination campaigns. As a result, the pharmaceutical chemist William Young and his new league, the London Society for the Abolition of Compulsory Vaccination (LSACV), managed to compete effectively with the NACVL for the control of the movement. Young established the LSACV in 1880, along with the merchant William Tebb and the bookseller William White. Under Tebb's guidance, the LSACV tried to recruit influential members of the Parliament and doctors in support of their campaigns. The LSACV published a handbill (Figure 4.3) to protest against compulsory smallpox vaccination for children, which illustrated their concerns. Its title was "Death the Vaccinator." It shows a skeleton vaccinating a young boy in his mother's arms while a police constable looks on. The image is titled "Compulsory Vaccination Act," and it is obviously intended to show its consequences. The police constable seems to oblige the mother to accept the risks of vaccination, which included death (hence "death the vaccinator"), while holding what is presumably a copy of the vaccination act. There is also a word game, as it is stated that vaccination can result in disease, but "generation" is replaced with "Jenner-ation" in an explicit attempt to refer to Edward Jenner who popularized vaccination.

The bill clearly expressed concerns about the dangers and the inefficacy of vaccination. First came the danger: "The matter used for vaccination started from a diseased beast, and has passed through innumerable children, taking up in its course no one knows what defilement, possibly the worst defilement, that will blast the existence of those who receive it." Then came the ineffectiveness: "And what are children vaccinated for? To keep off small-pox. But vaccination does not keep off small-pox. Every small-pox hospital contains vaccinated patients, who suffer and die like the unvaccinated." But children could not be vaccinated if parents did not give their consent. The advice given in the handbill was that the fine, if it was ever given, was a better choice than endangering the life of one's child.

The rhetoric about individual liberties was central in the anti-vaccination campaign and influential. The contrast between personal liberties and governmental mandates remains a central theme in today's anti-vaccination movements. As we see in the subsequent chapters, besides some very rare exceptions, there are no concerns any more about the safety of vaccines among scientists. However, the question of individual liberty and bodily autonomy remains. Shouldn't people have the liberty to make choices for

Figure 4.3 The handbill "Death the Vaccinator," published by the London Society for the Abolition of Compulsory Vaccination in the 1880s. It begins with the statement "What do we have we here? Death the Vaccinator with the policeman to assist him. Are children killed by vaccination? Who can doubt it?" Reproduced with permission from Bodleian Libraries, University of Oxford. John Johnson Collection: Public Services folder 4.

their own bodies and the bodies of their children? Why accept a medical intervention such as vaccination when one does not consider it necessary or acceptable? Do governments have the moral and legal right to impose vaccination, either by making it compulsory or by requiring a vaccination certificate for schooling and other activities? Isn't it a matter of individual liberty to make decisions for one's own body? It is to these questions that we now turn.

Individual liberty or collective responsibility?

As we saw, an important kind of argument against vaccination is that vaccine mandates conflict with ideas of bodily autonomy. I will not delve too deeply into the reasons why some people may refuse vaccination, because one way or another, all end up being about who decides what happens to one's body. We should also keep in mind that many of the reactions against vaccination are not against the scientific recommendation to get vaccinated itself. Rather, the reactions have been against the political decisions about making vaccination compulsory by imposing financial or social burdens on people refusing to get vaccinated, such as fines or school exemptions, respectively.

One set of reasons that people refuse vaccines is quite similar to those we saw in earlier sections of this chapter, which relate to notions of body purity. These people consider vaccines and vaccination to be incompatible with their worldviews and religious beliefs. Such claims have referred to the contents of vaccines as being of a kind that particular religions do not allow the use or consumption of. For instance, some vaccines contain porcine gelatin as a stabilizer, which is incompatible with the religious dietary restrictions of some Jews and Muslims. Or vaccine development can be based on cell lines from aborted fetuses, which some Christians find unacceptable. Details notwithstanding, some people perceive a conflict between being vaccinated and their religious beliefs. Philosopher Mark Navin has argued that such purity ideals are practically unrealizable, as we unintentionally inhale and ingest a lot we would rather not. In addition, such purity ideals are sometimes objectionable, as they do not morally justify the exposure of unvaccinated infants to infectious disease.[13] I tend to agree, but what I am concerned about here is not whether these people are right or wrong. My concern is whether their feeling that they should have the liberty to make their individual choices and follow the values they have chosen to live by conflicts with the responsibilities we all have as members of the communities in which we live.

As we saw in Chapter 2, vaccination is not a medical intervention intended for individual immunity only. It is also a means for generating and maintaining herd immunity, and thus vaccination is not just a personal issue. Therefore, the question that we need to answer is whether we should prioritize individual liberty or the collective good. Legal scholar Robert Field and bioethicist Arthur Caplan have argued that one can think of this dilemma as a contrast between competing ethical values. On the one hand, autonomy is the right of individuals to decide their own behavior. To exercise this right, it is necessary to be free from outside influences and limitations, as well as to be able to understand the consequences of one's decision and the available alternatives. On the other hand, the government also has the right to exercise coercive authority that violates the right to autonomy in the interest of promoting a different value that is usually about the common good such as protecting public health. To achieve this, advocates rely on the values of beneficence (acting for the benefit of others, for instance by preventing harm, or helping those in danger), utilitarianism (balancing the relevant factors to determine the best outcome for most, regardless of competing individual needs), justice (the fair, equitable, and appropriate distribution of goods that are not readily available to all), and nonmaleficence (acting against inflicting harm on others). In the debates about vaccines, both proponents and opponents accept these values but disagree over which one is more important when they are found to be in conflict.[14]

Among these, Field and Caplan have argued, the most prominent and longstanding argument put forth by mandate proponents is utilitarianism, which supports mandating a vaccine in order to prevent the harm to society that could be caused by unvaccinated individuals. Generally speaking, utilitarianism is the view that the morally right action is the one that produces the most beneficial outcomes for all. On this view, we ought to maximize the overall good, that is, consider the good of others as well as our own good, with the latter counting as much as anyone else's good.[15] For extremely contagious diseases, such as measles (see Chapter 10), even a small number of unvaccinated individuals in a local community can have a devastating impact. According to utilitarianism, mandating vaccines for everyone confers the greatest medical benefit for the greatest number of people, regardless of whether such a mandate is in the best interests of everyone or whether it ignores the right to autonomy. This is why utilitarianism, Field and Caplan have argued, in the interest of controlling disease spread, is the most compelling ethical consideration that competes with autonomy. As soon as the threshold for herd immunity is reached, it is in the interest of the community to coerce widespread vaccination.[16] In short, utilitarianism suggests that the fact that vaccination promotes public health is not just one moral reason to get vaccinated, but the most important moral reason. This entails that according to utilitarianism, anyone who refuses vaccination and thus does not contribute to the ultimate goal of public health would be acting immorally.

Not everyone agrees though. Philosopher Mark Navin has argued that utilitarianism has some odd consequences. According to utilitarianism, those parents who refuse to vaccinate their children because they are concerned about vaccine safety are perceived as acting immorally because they put the good of their own children above the common good. Utilitarianism entails that parents ought not to care about their children more than about the children of others, and other people more broadly. But this is odd. Navin argues, and I agree, that parents should be entitled to prioritize the well-being of their own children. So, in principle, those parents who refuse vaccination because of concerns about their children's safety should have the right to do so. But this in no way entails that these parents are right about vaccine safety, or that their vaccine refusal is ethically permitted. In fact, Navin has suggested, these parents have a moral obligation to vaccinate their children, but the reason is other than what utilitarian ethics suggest.[17]

According to Navin, parents have a moral duty of fairness to have their children vaccinated because herd immunity is a public good. A public good is anything that is available to anyone at no extra cost, and that is impossible to prevent people from having. For instance, when a country has a clean, unpolluted atmosphere, it is available to all once it exists and there is no way of preventing people from having it. Same is the case for herd immunity. Once it exists, it exists for everyone, both susceptible and nonsusceptible people. A public good is the outcome of social cooperation, and thus the responsibility for creating and maintaining it falls equally to all members of the community. This is done by establishing fair schemes by which people contribute equally to the intended public good. With respect to herd immunity, in order for it to be maintained, it is necessary that all new members of the community be vaccinated against particular infectious diseases. If they do not, there is the risk of losing herd immunity. Even if this does not happen, these people would be behaving immorally because they would be profiting from something to which they did not contribute, and which others have achieved for them. Therefore, people have a moral duty of fairness to get vaccinated, in order to maintain a public good, herd immunity, and in order to refrain from unfairly "free riding" on what the others have contributed. It should be noted though that a fair contribution should impose a reasonable and fair cost among those who contribute. Therefore, those people who are at high risk of vaccine complications do not have a moral duty of fairness to contribute to herd immunity—and it is primarily these people who are protected by it.[18]

Philosopher Alberto Giubilini has developed a slightly different argument, but still focusing on fairness. If we accept that herd immunity is the goal of vaccination policies, then there are ethical reasons to implement the least restrictive alternative to achieve this goal (where restrictiveness is measured in terms of the degree of limitation of individual autonomy). In other words, it is ethically imperative not to coerce everyone into vaccinating in order to achieve herd immunity if fewer people are sufficient for doing so. From this point of view, compulsory vaccination can be accepted only as a measure of last resort, when there is no better alternative, and only if it allows for conscientious objections to vaccination, thus limiting the autonomy of as few people as possible. Giubilini has argued that the less restrictive policies are preferable to more restrictive ones if the aim is achieving herd immunity. However, he has also suggested that the appropriate aim of vaccination policies is not herd immunity itself, but a fair distribution of the burdens entailed by herd immunity.

Giubilini has also questioned the various arguments that people ought to get vaccinated in order to refrain from causing harm to others and to maintain the public good of herd immunity. In his view, each individual only makes a very minor, actually negligible, contribution towards the public good of herd immunity. Therefore, it is difficult to argue that each individual should have an obligation to get vaccinated. This is actually an intuitive thought of many people. In the same way that many people refrain from voting in elections because they think that their single vote will not change much—and indeed on its own, alone, it does not—they may also think that not being vaccinated will not threaten herd immunity—and indeed it will not if everyone else is vaccinated.

But then how is compulsory vaccination morally justified? Giubilini has argued that there indeed exists an ethical principle that justifies an individual (moral or legal) obligation to be vaccinated: It is the principle of fairness. Not only is fairness relevant to policymaking as this is in large part what policies are about, but there are also no moral arguments against it as it is something that most people would agree to with respect to the distribution of certain burdens. Fairness should be perceived as equity rather than equality; this entails that it is not expected that everyone is to make the same contribution (equality) but a contribution that depends on a person's capacities (for instance, some people cannot get vaccinated because their immune system is compromised and so they should be exempted). Fairness in this sense could be implemented in the same way it is with taxation, where the taxes that people pay are not the same but depend on their income and fortune.[19] In this view, fairness is a more important principle than individual liberty when it comes to vaccination policies.

The discussion about the ethics of vaccination is just one instance of the broader question about the relation between science and ethics. Even if scientists all agreed about the safety and the efficacy of vaccines, there are ethical questions to be considered with respect to whether vaccination is voluntary or compulsory. The latter is a decision that cannot be made on scientific grounds alone, as one has to consider ethics as well. The conclusion reached that fairness is a more important principle than individual liberty when it comes to vaccination policies is not a scientific one. The decision on whether vaccination should be made compulsory or not can be informed by science (there is no point in considering making vaccination compulsory if it is not widely agreed that the vaccines under discussion are safe and effective), but it also requires more than science. Therefore, this is a decision that cannot be made based on science only.

When do human life and personhood begin?

Let us consider another example, where science can inform decisions but cannot alone support them: the termination of pregnancy, or abortion, which combines scientific, ethical, and legal perspectives to the question, "when does human life (and personhood) begin?" To delve into this question, let us first very briefly consider the development of the human embryo. Human reproduction is possible when a sperm cell and an ovum fuse during the process described as fertilization. Whereas each woman usually produces one ovum per menstrual cycle, a man can produce around 40,000,000 sperm cells per ejaculation. From these, only a few hundred will reach the ovum. One of these sperm cells will fuse with the ovum, and this is the phenomenon we describe as *fertilization*. Once this happens, a reaction is initiated within the ovum that blocks the entry of other sperm cells. The fertilized ovum will begin dividing and within a week or so will be transformed into a blastocyst, a spherical cell mass of approximately 150 cells. It is in this form that the early human embryo can be implanted in the uterus of the mother (*implantation*). This is when pregnancy begins. The human embryo will continue to grow in cell number while its cells differentiate and give rise to the different cell types. A key stage of this process is *gastrulation*, which occurs during the third week after fertilization. At this stage, a massive reorganization of the cells takes place, resulting in three different types of cell layers: the endoderm from which the epithelial tissues and glands of the digestive and respiratory systems are derived, the mesoderm from which connective tissues (cartilage, bone, blood) and muscles (cardiac, skeletal, smooth) are derived, and the ectoderm from which skin and parts of the nervous system are derived. The resulting multilayered and multidimensional structure is called the gastrula.[20]

When does human life begin during this developmental process? In fertilization? In implantation? In gastrulation? Or even at some later point in time? According to legal scholar Steven Andrew Jacobs, the vast majority of scientists believe that life begins at fertilization. Jacobs analyzed the results of a survey among 5,577 biologists from 1,058 academic institutions around the world and found that 5,337 of them (96%) agreed with the statement that human life begins at fertilization. Based on the results of his survey, Jacobs concluded: "Given the size and breadth of the international sample of biologists in the present study, its results represent strong support for the claim that there is a scientific consensus on the view that a human's life begins at fertilization." He cautiously added that one survey is not enough and that further studies are required "… to replicate these findings." And he concluded that "If this study's findings are confirmed, then the fertilization view can be promoted by scientists and shared with members of the public to ensure they are informed on the biological perspective on when a human's life begins, as this would empower them to make informed reproductive decisions."[21]

In contrast, developmental biologist Scott Gilbert has argued that very different views have been expressed by biologists, and that there is no consensus among them as to when human life begins. He has suggested that scientists whose expertise is the study of embryos hold one of the following views about when human life begins, which also corresponds to when an entity becomes a person (personhood):

- *Fertilization*: In this view, personhood begins at fertilization, when the zygote comes to exist, wherein the DNA from two parents combines to form a new, unique genome.
- *Gastrulation*: In this view, personhood begins when the embryo completes gastrulation, because after this stage it is no longer possible for twins to emerge and thus each embryo will give rise to only one individual.
- *Acquisition of the human electroencephalogram (EEG) pattern*: In this view, it is the functioning of our central nervous system that determines when personhood begins. As the loss of the human EEG pattern usually determines the end of life (cerebral death, even though one's heart may still be functioning), it is argued that in the same sense it is

the acquisition of the human EEG pattern around 20–24 weeks after fertilization that marks the point at which human life begins.

- *Birth*: In this view, personhood begins at or around birth, because a fetus should be considered an individual only when it can survive outside the body of its mother. This is usually determined by the respiratory system, as a fetus can survive outside the uterus only when its lungs are sufficiently mature, usually around 28–34 weeks of pregnancy.

Finally, there exist scientists who claim that the acquisition of personhood is gradual or that the question of when life and personhood begin is not a biological one. Gilbert's overarching conclusion was that "There is no consensus among biologists as to when personhood begins."[22]

What is the cause of the entirely opposite conclusions reached by Jacobs and Gilbert? And who should we listen to? At first sight, this is a comparison between the views of a few thousand biologists surveyed by Jacobs and of a handful of biologists quoted by Gilbert. Should we trust the view of the many or of the few? You might be tempted to argue that the larger sample would provide more reliable results. But it is not necessarily so. As I explain in detail in Chapter 6, what matters when it comes to scientific conclusions is the consensus view of expert scientists, that is, what scientists who are experts in a particular domain, not just any person with a scientific degree, think about an issue. And this is where we find a big difference between Jacobs and Gilbert. Even though Gilbert did not conduct an empirical study surveying experts, he cited dozens of papers in expert journals related to pediatrics, obstetrics and gynecology, and genetics. As he wrote, he was interested in the views "... held by different groups of biologists, especially those scientists whose profession is the study of embryos."[23] In short, Gilbert was interested in the views of experts. In contrast, Jacobs gave no detailed information about the biologists who participated in his survey. He mentioned that the vast majority of them had a PhD, but not what kind of PhD. Lumping together biologists with a variety of expertise poses questions about the validity of the findings. Are the views about embryo development of a plant biologist, a fly geneticist, an insect ecologist, or a biology educator (this is me) as valid as those of a mammal developmental biologist? I respectfully believe they are not (see Chapter 6 and Table 6.1 for the criteria for identifying experts).

But this is not all. Jacobs had initially contacted 62,469 academic biologists, of whom 7,402 participated in the online survey and 5,577 provided data for analysis. More information about the demographics of this sample should have been provided to ensure that this was not a biased sample. As philosopher and biologist Sahotra Sarkar put it, "That result is not a proper survey method and does not carry any statistical or scientific weight. It is like asking 100 people about their favorite sport, finding out that only the 37 football fans bothered to answer, and declaring that 100% of Americans love football."[24] Finally, if you read Jacobs' paper closely, the 96% stated agreement is likely an artifact of the analysis. This is the percentage of people who were consistent across all questions, being for or against the view that life begins at fertilization: "... (1) those who either affirmed each statement (Q1-Q5) and wrote about the fertilization view in response to the essay question (Q6), and (2) those who rejected each statement and wrote about some later point in development."[25] But what Jacobs did not mention was how many these people were. If only 100 people were consistent, and 96 of them thought that life begins at fertilization, this does not tell us much about the whole sample. It is also worth noting that when participants were asked to answer the open-ended question (Q6) "From a biological perspective, how would you answer the question 'When does a human's life begin?'," Jacobs reported that "68% of biologists (1898 out of 2793) represented the fertilization view in response to Q6's open-ended essay question." What happened to the 96%? It seems that the results of the Jacobs survey should be considered with caution.

I personally find the conceptual variation described by Gilbert much more plausible and much more likely to be true than the results of the Jacobs survey. In my view, there is a simple reason that there is no consensus among scientists about when personhood begins: This is not a question that can be answered only on scientific grounds. It is certainly a question that can be informed by science, considering what happens during embryonic development. But scientific knowledge alone cannot tell us when an entity has moral or legal rights. There are signs of the phenomenon we call life in all stages: the sperm and the ovum, the zygote, the blastocyst, the implanted embryo, the fetus, and the born individual, all exhibit features of the biological phenomenon we call life. But human life is not restricted to its biology only. There are other aspects that are important to consider too. Here are a few things to consider: What about those fertilized embryos that are naturally not implanted? What about those fertilized embryos that are implanted on sites other than the uterus and do not develop? What about those implanted embryos that are accidentally aborted? One can decide that one or the other stage of development is more important than the others and consider it as marking the beginning of personhood. But such a decision would be subjective and arbitrary as there is no simple scientific justification for why it should be the case.

What is crucial to understand? (and also teach in schools!)

In the previous section, I argued that the questions of whether we should prioritize individual liberty or collective responsibility, or about when human life and personhood begin, are not scientific ones. As historian Naomi Oreskes nicely put it: "Distinctions between the scientific and the social matter, because they rightly affect our choices, and because they help us to distinguish between arguments that may be persuasive to our audiences and arguments that are doomed to fail because they don't address their underlying concerns."[26] However, scientific knowledge helps set these questions in a helpful context.

The question of when human life and personhood begin clearly shows that taking science into account is important but insufficient to provide a complete answer. From a biological point of view, the sperm cell and the ovum, the zygote, the blastocyst, the gastrula, the fetus, and the newborn are equally alive. All these structures consist of one or more cells that exhibit the typical features that we describe as "life." Of course, some of these are much larger and much more complex than others. But they all have metabolism, DNA expression, protein synthesis, and other cellular activities. The phenomena of life at the level of the gastrula, including cell division and differentiation, are more complex than those in the ovum. But the basic features of life occur in all cells. This is what biology can tell us. What it cannot tell us is which of these different stages marks the beginning of the life of a human with moral rights and obligations. The answer will mostly depend on one's worldviews, and this is why a variety of views among expert developmental biologists was reported by Gilbert. I personally think that there is no reason for considering the beginning of human life and personhood before the implantation of the embryo in the uterus of the mother. Once this happens, the most likely outcome is birth, even though there are many exceptions. But given that it is natural for many early embryos not to implant in the uterus, or to be implanted but subsequently be spontaneously aborted, the implantation of the embryo in the uterus seems to be a turning point for the beginning of a new life.[27]

It is difficult to estimate how many early embryos naturally fail to implant in the uterus and how many implanted embryos are naturally lost for various reasons. The latter process is referred to as a miscarriage. This term is used to refer to all pregnancy losses from the time of conception until 24 weeks of pregnancy, and it seems that about 50% of women experience one or more miscarriages during their lives.[28] It also seems that the risk of miscarriage rises with the mother's age. A Canadian study examined 94,346 pregnancies between 1961–1974 and 1978–1993, finding that women over 35 years old had a notably higher rate of fetal death than younger women. Specifically, women aged 35–39 years were about twice as likely to experience fetal death compared to 30-year-olds, and this risk was even greater for women over 40 years old.[29] Another study in Denmark examined 1,221,546 pregnancies among 634,272 women from 1978 to 1992. It was concluded that the risk of spontaneous abortion increased from 8.9% for women aged 20–24 to 74.7% for women aged 45 and older.[30]

It has long been established that changes in chromosome number or structure, which are described as chromosome abnormalities, are linked to miscarriages. A review of 13 studies, involving over 7,000 women who had a single miscarriage, found chromosome abnormalities in almost half of the analyzed samples. The most common abnormalities were trisomies, where an additional chromosome results in a total of 47 chromosomes, and they were found in about 6 out of 10 of the abnormal embryos. Similarly, six other studies with over 1,000 women who had multiple miscarriages found chromosome abnormalities in 39% of the samples, with trisomies present in 65% of the abnormal embryos.[31] Such changes that result in an additional chromosome are described as aneuploidies. Miscarriages caused by aneuploidies are relatively common in the later stages of pregnancy, affecting about 5% of pregnancies. However, there is evidence that aneuploidies are even more prevalent before implantation. In one study, researchers investigated the prevalence of aneuploidies in preimplantation embryos, obtained from couples undergoing in vitro fertilization. In particular, the researchers studied 420 fertilized ova, 738 embryos at the stage of 8–16 cells (also called cleavage stage), and 1,046 blastocysts. The results showed a high frequency of chromosome abnormalities in embryos before implantation: 74% of the fertilized ova, 82% of the cleavage embryos, and 58% of the blastocysts studied. For all three groups, the observed abnormality rate was higher for older women.[32]

If these findings are an indication of what happens in nature, then it seems that many of the embryos emerging from fertilization do not result in a pregnancy. There are many assumptions and limitations in the studies just considered, but they also clearly show that it is not a given that fertilization will naturally result in a new human life. Hence my doubt that human life and personhood could be considered to begin before implantation. Having said that, I do not argue that I am right and that others are wrong. I just state my view to show how scientific evidence can inform a moral decision that is mostly subjective. Someone else might interpret these findings in a different way I do. There is no problem with subjective decisions insofar as they only affect the person who decides and nobody else. For instance, there exist diseases that are due to pathogens transmitted from person to person, and others that are not. For instance, tetanus is due to bacteria usually found in soil, dust, and manure, which can enter the human body through cuts or puncture wounds caused by contaminated objects. However, these bacteria are not transmitted from one person to another. Therefore, as tetanus is not a contagious disease and as people cannot be infected by other people, choosing not to get a tetanus vaccine *is* indeed a personal choice. It affects only the person who decides not to get vaccinated, and nobody else around them. But this is not the case for contagious diseases such as measles.

We now know that vaccines have significantly decreased morbidity and mortality from infectious diseases. In 1974, the World Health Organization launched the Expanded Programme on Immunization (EPI) in order to make vaccines available to people all over the world. A recent study aimed to quantify the public health impact of vaccination globally in order to mark the 50th anniversary since the EPI's inception. It has been estimated that since 1974, vaccination has averted 154 million deaths. These include 146 million children younger than five years old, among whom 101 million are infants younger than one year old. The researchers estimated that vaccination has accounted for 40% of the observed decline in global infant mortality, and 52% in the African region. In 2024, a vaccinated child younger than 10 years old is 40% more likely to survive to their next birthday relative to a hypothetical scenario of no vaccination.[33]

I am inclined to think that besides our moral duty to contribute to a public good, we can also think about vaccination in more pragmatic terms by way of analogy with taxation. I do not believe that there are any people who are happy to pay taxes, but taxation is essential for funding public goods and services that are necessary for societal functioning and well-being. These public goods include infrastructure (roads, bridges, and public transportation systems), public utilities (access to clean water, electricity, and sanitation services), education (free, accessible, and quality schools), security (maintaining law and order, and ensuring the safety of citizens through police, emergency services, and national defense), justice (ensuring fair and impartial administration of justice through courts), social welfare (offering unemployment benefits, housing assistance, and food support for those in need), environmental protection (safeguarding the environment through regulations, conservation efforts, and initiatives to address climate change and pollution), and last but not least healthcare (providing hospital treatment whenever necessary). Whatever is around us and exists thanks to the public sector, it is supported by the taxes we all pay. This is why we all have both a moral and a legal obligation to pay taxes. Comparing vaccination to taxation may at first sound odd, as the risks and the morals involved are different. However, they are both based on collective action towards public good provision. The difference is that the grounds are not moral but pragmatic. If we like the public goods we are offered, then it is also beneficial for us to contribute in order to maintain them. But besides this pragmatic argument, there is also a moral one.

In some sense, imposing taxation goes against individual liberty, according to which people should be free to decide what to do with the money they legitimately earn. For some people, taxes may even do harm as they deprive them of money that they might use to improve their living conditions. So, whereas the money one earns may not even be sufficient for a comfortable life (however defined), one is obliged to give some portion of their income to the government in the form of taxes. You can imagine that public servants might not have a problem returning back to the government some of the money they earned that all came from the taxes of other people. But how about those people who have their own businesses, and who struggle to sustain them every day? Who knows whether they will earn anything tomorrow? Is there a moral justification for the government to demand a part of their earnings? Even if this is done in an equitable manner, meaning that richer people pay more and poorer people pay less, or that those who work in the private sector have some kind of tax relief to compensate for their insecurity, what is the moral justification for the government imposing taxation on people's legitimate earnings?

The answer (being aware that things are actually far more complicated than that[34]) could be that there is a moral justification for the government's imposition of taxation on people's legitimate earnings in order to maintain all kinds of public goods that governments are expected to offer. If one could live without these, then one might have a legitimate argument against paying taxes. But if we cannot imagine life without these public goods and if we take them for granted, then we need to remember that these exist thanks to the contributions that we all make with our taxes. It is in a similar sense that herd immunity is a public good to which we all contribute by being vaccinated. Actually, taxation can be more harmful (in the sense of affecting everyday life) than vaccination, as those who cannot afford paying taxes are many more than those experiencing side effects from vaccines. One might argue here that taxation does not interact with our body, and that there is no risk of dying when we pay taxes. The answer would be that if one looks at the data, the risks of dying from vaccine side effects are extremely rare, and as mentioned vaccines have reduced the deaths from infectious diseases enormously.

So, what is there to do? Based on in-depth interviews with parents in the USA, sociologist Jennifer Reich has concluded that refusing or delaying vaccination are choices rooted in an individualist parenting ideology that prioritizes parents' right to choose a personalized healthcare plan for their children, rather than accepting the moral obligations of public health. Many parents prefer an individualized approach to vaccination, Reich explained, with arguments like: "every child has a unique immune system, so why are the vaccine schedules all the same?" Those parents' choice, therefore, has been to ask their doctors to customize vaccine schedules based on their own preferences, so that they feel they are getting personalized care. The perception that every child is unique has broader implications, for instance, when schools are expected to adapt to individual children's needs rather than expecting children to adapt to school. As Reich noted: "Individualism has come to define what it means to raise children, and vaccines are not exempt from this way of thinking."[35]

Reich also told the story of the mother of three unvaccinated boys who argued that if she had a girl, she would have probably had her vaccinated against measles before adolescence. The reason was her awareness that measles could be devastating for fetal development, were her daughter to get pregnant at some point. But she was not worried about her boys with respect to getting measles. Rather, she was concerned about mumps because it could cause sterility to them. So, she wanted them to contract mumps before adolescence, and she would have them vaccinated if necessary. Whether or not this mother's judgments about the particular vaccines are correct is not the issue here. The issue is that this mother only considered her sons as individuals, and not as members of a community. She did not consider that even if her sons did not have any problem themselves when contracting measles, they might be around pregnant women whose fetuses—in her own logic—would then be in danger. Therefore, her sons might be responsible for infecting a pregnant mother and thus badly affecting the development of her unborn child.[36] Being unvaccinated entails a risk both for oneself and for the others. We vaccinate ourselves and our children both for our individual protection, and for protecting others.

Therefore, governments and authorities need to refrain from promoting the individual benefits of vaccination only and ensure the promotion of both the individual and the collective benefits. A study with more than 2,000 participants from South Korea, India, Vietnam, Hong Kong, the United States,

Germany, and the Netherlands has shown that this can work well. Participants were divided into three groups that received different information about herd immunity: One group was in a condition where the social benefit of vaccination was emphasized: "... when you get vaccinated, then you can protect others who are not vaccinated"; a second group was in a condition where the individual benefit was emphasized "... the more people who are vaccinated in your environment, the more likely you are protected without vaccination"; and a control group did not receive any information about herd immunity. The groups were then given two scenarios, one for a highly contagious and one for a less contagious fictitious disease. In each of the two scenarios, the participants read about the disease, the respective vaccine, and the probability of vaccine adverse events. It was found that explaining the concept of herd immunity led to higher vaccination intentions compared to the control group, particularly for western countries, which are considered as more individualistic compared to the eastern ones. Therefore, effectively communicating the concept of herd immunity might help raise vaccination levels.[37]

I believe that the stories in this chapter clearly show that there are questions that science alone cannot answer. How we behave as members of a society, what our moral obligations are, when our personhood begins, when we have moral rights, and so on are questions that can be informed by science, but cannot be answered by it.

5

"The practice of medicine is based on calculated risk"

Why uncertainty is inherent in science

Paralyzed by fear

> Good evening, Mr. and Mrs. America, and all the ships at sea ... Attention all doctors and families: the National Foundation for Infantile Paralysis plans to inoculate one million children with a new vaccine this month. ... The U.S. Public Health Service tested ten batches. ... They have found (I am told) that seven of the ten contained live (not dead) poliovirus. ... That it killed several monkeys. ... The name of the vaccine is the Salk vaccine; named for Dr Jonas Salk of the University of Pittsburgh.[1]

It is April 4, 1954, and this is Walter Winchell's voice on his popular Sunday night radio show.[2] Winchell is considered the inventor of the gossip column, but in this case what he reported was not gossip at all. A story published in the November 23, 1953 issue of *Time* magazine reported concerns about the safety of Salk's vaccine against the poliovirus, the virus causing poliomyelitis (hereafter polio). The concerns were raised by Albert Milzer and his collaborators at Michael Reese Hospital in Chicago, because they could detect active virus in the vaccine they had produced via the same procedure followed by Salk himself.[3] Then in January 1954, Milzer and his collaborators published a study that proposed a different method for producing a polio vaccine that was virus-free. Nevertheless, their article did not include any criticism of Salk's method, but only proposed an alternative. They noted though "... that very critical safety tests for complete inactivation were essential because of the questionable safety of chemically inactivated poliomyelitis vaccines studied in the past."[4]

What caused Winchell's reaction in April? It may not be a coincidence that just a few days earlier, on March 29, 1954, Salk's portrait was on the cover of

Trusting Science. Kostas Kampourakis, Oxford University Press. © Oxford University Press (2025).
DOI: 10.1093/oso/9780197787106.003.0005

Time magazine, with the caption "POLIO FIGHTER SALK Is this the year?" The cover story went like this:

> This spring. Dr. Salk's vision and his delicate laboratory procedures and logarith-
> mic calculations are to be put to the test. Beginning next month in the South and
> working North ahead of the polio season, the vaccine that Salk has devised and
> concocted will be shot into the arms of 500,000 to 1,000,000 youngsters in the first,
> second and third grades in nearly 200 chosen test areas. A few months after the 1954
> polio season is over, statisticians will dredge from a mountain of records an answer
> to the question: Does the Salk vaccine give effective protection against polio?[5]

As many children were expected to receive a vaccine that might not be entirely safe, perhaps Winchell thought that it was his duty to inform society; or he was simply looking for a catchy story for his show. Whatever the case, it is estimated that about 150,000 children were not vaccinated as a result of Winchell's program and the public discussion following that.[6]

In the 1950s, polio was the most feared issue after the atomic bomb. Yet, it was not the most lethal disease; in fact, it affected very few people compared to other infectious diseases such as measles. It is estimated that paralysis develops in less than 1 in 200 people infected; among 100 people who develop paralysis, less than 10 die; whereas most of those infected do not have any visible symptoms.[7] However, because poliomyelitis mostly affected children, even infants, it had a devastating impact on everyday life. In some cases, infants, even newborns, were obliged to be inside "iron lungs," with only their heads left outside the chamber. Pumps raised and lowered the pressure inside the chamber. When the pressure fell below that of the lungs, outside air filled the lungs; when it rose above that of the lungs, air was expelled from the lungs. For many people, this was their only way to breathe. The reason was that their diaphragm, the muscle that supports our breathing, was paralyzed. In other cases, children were completely paralyzed and were unable to move on their own. Finding a vaccine against it seemed like the only way to go.

Already by 1909, Karl Landsteiner in Vienna and Simon Flexner in New York had managed to show that polio was caused by a virus (we now know that it is an RNA virus; see Box 2.1). In 1916, a polio epidemic in New York City that resulted in the death of 3,310 people brought this disease to the center of attention. But it took several years before polio vaccines that were tested on humans were produced. It was in 1935 that the results of tests of two kinds of polio vaccines, developed independently, were presented: Maurice Brodie in New York had developed a vaccine with a "killed" virus, whereas

John Kolmer, working in Philadelphia, had developed a vaccine with a "live attenuated" virus. The differences between these two types of vaccines are crucial for the story in this chapter (see also Box 2.2). The "killed" virus in the vaccines cannot proliferate, but its molecules are recognized by our immune system, resulting in an immune response and later in immunity. However, adjuvants may be needed to boost the immune response, and usually more than one dose of the vaccine is required. In contrast, the vaccines with a "live attenuated virus" contain a weakened but live virus. Its recognition by our immune system is similar to the one that happens during natural infection, and so the immune response in this case is stronger and more long-term than the one caused by the "killed virus." However, the "attenuated live" virus has the potential to proliferate, and eventually cause disease. This is why extensive testing is required before such a vaccine is administered, whereas it may not be appropriate for people with compromised immune systems.[8]

Kolmer reported that his vaccine with a "live attenuated" virus was successful. Among 42 monkeys receiving large doses of the vaccine, none had shown any signs of infection by polio, whereas in a control series of 20 monkeys receiving similar doses of the virus, one developed paralysis. From these results, Kolmer inferred that the infectivity of the virus had been reduced thanks to the method of preparation of the vaccine. He also reported that it had not been possible to immunize monkeys with vaccines containing a virus "killed" by heat or chemicals such as formalin. These results made Kolmer believe that for vaccination to be effective, some form of active virus should be included in the vaccine. These results also made him sufficiently confident about the vaccine to administer it to a small number of individuals, among them children, with no ill effects except for some local reactions, and eventually to 10,725 individuals. As already explained in Chapter 2, to evaluate the impact of a vaccine, a control group is required. Kolmer considered that the unvaccinated individuals in the vicinity of the homes of vaccinated people could serve as controls. However, the results were not that good, as poliomyelitis developed in 10 people who had received one or two doses of the vaccine. Kolmer argued that "In at least some of these, it is a reasonable assumption that the vaccine was not responsible for the attacks, and in no instance has poliomyelitis developed after the full 3 doses of vaccine had been given."[9]

These conclusions were not left unaddressed. Thomas Rivers, a prominent virologist at the Rockefeller Institute, expressed his concern in an article in the same journal issue about the lack of a "... comparable number of unvaccinated children, properly chosen in regard to location and age, to act as controls for the efficacy of the vaccine in the individuals receiving it." He also questioned Kolmer's conclusion that the 10 cases of paralysis were not due to the vaccine, given that they did not occur in people who had received three doses of the vaccine, but in those who had received one or two doses. Most importantly, finally, Rivers noted that the vast majority of the vaccinated children would never have contracted poliomyelitis.[10] It had become clear by that time that only a few people infected by the virus developed paralysis, so a badly prepared vaccine could be more dangerous than the virus itself.

Brodie was more cautious in his work, which he did under the supervision of bacteriologist William Park. He was aware from animal experiments that the use of live virus resulted in the most effective vaccine, but also the most dangerous one. He thus decided that a "killed" virus was the safest way to go. Experiments with monkeys were encouraging, as they had shown that only a few of those who had received less than "the estimated infective dose of the virus" had developed the disease. Thus, human vaccinations came next with 7,000 vaccinated people living in areas where they were likely exposed to the poliovirus. These were compared to 4,500 unvaccinated people living in exposed areas, with Brodie noting that "In so far as possible each vaccinated individual was matched with a control in the same district and of the same age." The results presented were considered as encouraging given that among those 7,000 people exposed to the virus, none had developed the disease, compared to 5 in the control group who did. However, in a footnote, Brodie wrote: "We have recently learned of a case developing 13 days after a single dose. A similar second case has, only been reported, but the diagnosis is not certain."[11] Rivers commented on Brodie's study too. His conclusion was that indeed no case of poliomyelitis could be ascribed to Brodie's vaccine. But he could neither find any evidence for or against its efficacy, predicting that in that form Brodie's vaccine would be "reasonably safe but ineffective."[12]

The reaction of the scientific community to the Kolmer and Brodie vaccines was very critical, to say the least. Towards the end of 1935, Simon Flexner, one of the pioneers of polio research, wrote in a commentary published in *Science*: "No adequate evidence has been presented showing that through the action of physical and chemical agents the virus of poliomyelitis may be attenuated so as to preserve its immunizing properties, while being deprived of its potential paralyzing power."[13] The virus was either destroyed and no longer had any immunizing properties or reduced in concentration and immunized some animals while it paralyzed others, from which a fully active virus could be found. In the 1930s, there was no serious prospect for a polio vaccine any time soon.

A calculated risk

Franklin Delano Roosevelt is known for serving as the 32nd president of the United States of America; for being the only US president to have served more than two terms in office, for 12 years from March 4, 1933 until April 12, 1945; and for leading the US during World War II. But what is perhaps less known is that he was paralyzed since 1921. The diagnosis was polio, even though this has been disputed. Nevertheless, Roosevelt was interested in the disease and in doing something about it. As a result, in 1938, he established the National Foundation for Infantile Paralysis (hereafter the Foundation), a fundraising and volunteer organization, and appointed his friend Basil O'Connor as its director. The Foundation launched an initiative that became known as *The March of Dimes*, which encouraged citizens to contribute their dimes (10-cent coins) to fund polio research. Millions of letters (and dimes) started arriving at the White House in 1938. Soon enough, there was money for research on a polio vaccine. But perhaps in part due to World War II, and in part due to the reactions against the Kolmer and Brodie vaccines, no serious attempt was made to develop one. Crucially, there were also scientific reasons for this delay.

As the poliovirus affects the nervous system, it is not surprising that experimentation with monkeys was the best means for developing a vaccine before administering it to humans. But monkeys were not as easily available as mice, and they were also expensive to buy and maintain. Worse than this, they had to be killed during the experiments. However, everything changed in 1941 when John Enders at Harvard University, with his collaborators Frederick Robbins and Thomas Weller, succeeded in developing cell cultures of the poliovirus, work for which they received the 1954 Nobel Prize in Physiology and Medicine. This was an important advance towards the development of the polio vaccine because with these cell cultures, it was possible to obtain large quantities of poliovirus for research and confirm its presence by simply observing cells without killing any animals.

The first person to attempt to develop a polio vaccine was Hilary Koprowski. Working in New York, he published his results about a vaccine with a live virus in 1952. Most experts at the time agreed that this was the best way to go. Enders, for instance, expressed in 1954 the view that "On the basis of the experience with the vaccines against smallpox and yellow fever, it is probable that the ideal immunizing agent against any virus infection should consist of a living agent exhibiting a degree of virulence so low that it may be inoculated without risk."[14] Similar was the view of another longtime polio researcher, Albert Sabin, who in 1953 noted that "Unquestionably, the ultimate goal for the prevention of poliomyelitis is immunization with 'living' avirulent virus which will confer immunity for many times or for life."[15] But there was also a scientist who thought otherwise. That was Jonas Salk.

Salk preferred developing a vaccine with a "killed" virus. In 1951, after having experimentally established that three types of polioviruses existed, he received funding from the Foundation to develop a vaccine against polio. In 1952, the worst ever polio outbreak took place, and this motivated the beginning of clinical trials for Salk's vaccine. The chair of the Foundation committee overseeing the trials was Thomas Rivers. In 1953, Salk presented to the Immunization Committee of the Foundation the results of his tests of a polio vaccine with a "killed" virus on children (these were children in the D. T. Watson Home for Crippled Children and the Polk State School for the Retarded and Feeble-Minded; it is sad to read that they used these children for such an experiment, but the ethics of the time were different from today). In the published study, Salk wrote: "Although the results obtained in these studies can be regarded as encouraging, they should not be interpreted to indicate that a practical vaccine is now at hand. However, it does appear that at least one course of further investigation is clear. It will now be necessary to precisely establish the limits within which the effects here described can be reproduced with certainty."[16] Salk was cautious, but as we saw, other researchers had failed to reproduce his results.

Box 5.1 Categories of clinical studies[17]

There are two main categories of clinical studies:

- *Experimental studies*, in which the researchers themselves assign the exposures (e.g., a vaccine or a drug). These can be:
  - randomized, in which the exposures are assigned by a truly random technique (with concealment of the upcoming assignment from those involved);
  - nonrandomized, in which another allocation scheme was used, such as alternate assignment.
- *Observational studies*, in which the researchers observe usual clinical practice. These can be:
  - analytical, which include a comparison (control) group and which can be further divided into the following categories:
    - cohort studies, which follow people forward in time from exposure to outcome;

> - case-control studies, which trace back from outcome to exposure;
> - cross-sectional studies, which are like a snapshot, measuring both exposure and outcome at one time point;
> o descriptive, in which there is no comparison group.

It was eventually decided to proceed with a large-scale clinical trial using two different methods in parallel (see Box 5.1). In total, 1,349,315 children (aged six to nine years) participated in the 1954 vaccinations. They were called the pioneers of polio. On the one hand, more than 945,000 children either received the vaccine (one-fourth of them) or received nothing and were just observed (three-fourth of them) (vaccine/observed controls). On the other hand, more than 400,000 children received either the vaccine (half of them) or the placebo (the other half)—without knowing which one (vaccine/placebo). Why these two methods? In principle, only the second method (vaccine/placebo) can provide reliable results, because it eliminates many kinds of biases as neither the doctors nor the children or their parents knew who had received what. However, many states did not accept this, as they wanted to offer the vaccine to any child who wanted to receive it. This is where the first method was implemented, with the nonvaccinated children serving as observed controls.[18] Salk himself was in fact a proponent of this approach, having written a complaint letter to Basil O'Conor a few months before: "The use of a placebo control, I am afraid, is a fetish of orthodoxy and would serve to create a 'beautiful epidemiologic' experiment but would make the humanitarian shudder and would make Hippocrates turn over in his grave." For Salk, there was no moral justification for intentionally injecting children with a placebo for scientifically evaluating the vaccine, instead of immunizing them against the disease.[19]

It was in fact Harry Weaver, the Foundation's director of research, who had insisted on the use of a double-blind, vaccine/placebo trial, as he wanted to have solid evidence for or against the effectiveness and the safety of the vaccine. He was aware that the members of the expert advisory committee such as Thomas Rivers would react if there was no such procedure, and he wanted to refrain from having the criticism by experts undermine the public confidence in the vaccine. But Weaver also wanted to proceed with the trials. In a memorandum of January 30, 1953 to Basil O'Connor after a meeting with Thomas Rivers, he had written:

The practice of medicine is based on calculated risk. Where the risk is known, the physician elects to follow the course that provides the greatest benefit with the least risk of incurring any untoward effects.

It is impossible at this stage of development to predict the degree of efficacy on the one hand, and the degree of safety, on the other, of the poliomyelitis vaccine that has been developed. These questions can only be determined after injecting relatively large numbers of human beings.

There is no question of the facts that with additional research: (1) A still more effective poliomyelitis vaccine could be produced; (2) We would be better informed as to the kind and frequency of untoward effects that might result from the use of the vaccine; and (3) We would be better informed with respect to the best route of inoculation, and the best time for administration, of the vaccine to obtain maximal protection against paralytic disease ...

If such research is carried out, a very considerable amount of time will elapse before a poliomyelitis vaccine is made available for widespread use; with the result that, in the interim, large numbers of human beings will develop poliomyelitis who might have been prevented from doing so had the vaccine been made available at an earlier date.[20]

Eventually, Weaver resigned under pressure before the trials began. The person directing the trials was Thomas Francis, a prominent virologist and past mentor of Salk. He managed to convince the Foundation's officials to accept both methods.

Hundreds of people were involved in the trials, and it took time to analyze the results that were announced in 1955: the vaccine was 80%–90% effective in preventing poliomyelitis (60%–70% against type I and more than 90% against types II and III). The overall results are presented in Table 5.1. Two observations are important. First, that in both the placebo and the observed areas, there were three to four times fewer paralytic cases among the vaccinated children than among the unvaccinated ones. Second, that this notwithstanding, the frequency of paralytic cases was always less than 10 in 10,000 or 1 in 1,000. Most remarkably, there seemed to be no cases of paralysis among the more than 420,000 vaccinated children. This was in part due to the strict quality control measures implemented, which required obtaining 11 consecutive batches of the vaccine that had passed the safety tests and reviewing the protocol for each batch. Francis and the Foundation were very serious about safety.

Unfortunately, this did not seem to be the case for the commercial vaccination program of 1955, which followed the trials, and which was overseen by the federal government. On April 25 and 26, 1955, William Workman who was in charge of the vaccination program was informed that six children who had received the vaccine produced by Cutter Laboratories had

Table 5.1 The results of the 1954 Salk vaccine trials (based on Francis T. Jr., Korns R., Voight R., Boisen M., Hemphill F., Napier J., et al. (1955). An evaluation of the 1954 poliomyelitis vaccine trials: summary report. *American Journal of Public Health*, *45*(suppl), 1–50)

	Placebo areas		Observed areas	
	Vaccinated	Placebo	Vaccinated	Observed
Number of children	200,745	201,229	221,988	725,173
Number of paralytic cases	33	115	38	330
Frequency of paralytic cases	1.6 in 10,000	5.7 in 10,000	1.7 in 10,000	4.5 in 10,000

a been paralyzed. At the time, it could have been possible that the Cutter vaccine was not the cause of polio because those children had received only one dose of the vaccine, the vaccine was anyway not 100% effective even after the three doses, and polio epidemics had occurred recently. However, already in 1954, Bernice Eddy, whose job was to test vaccines from five different companies on monkeys, had found that the inactivated vaccine manufactured by Cutter Laboratories contained live poliovirus, causing paralysis in monkeys. Eddy had reported her findings to Workman, who was her supervisor, but those were never shared with the Vaccine Licensing Advisory Committee. In addition, Julius Youngner, Salk's close collaborator, had visited Cutter Laboratories and had remarked that their processes of inactivating the virus were not effective. Youngner had informed Salk of his observations and offered to write a letter to inform Basil O'Conor. Salk replied that he was going to write it himself, but it seems that he never did. Unfortunately, because the inactivation of the virus in the Cutter vaccines was insufficient, live poliovirus was likely injected into 120,000 children; it is estimated that as a result, 220,000 people were infected, 70,000 developed muscle weakness, 164 were permanently paralyzed, and 10 died.

Despite this very serious incident, Salk's vaccine was considered to be safe and effective. It was given as a shot and thus induced immunity in the bloodstream, which prevented polio from traveling to the brain and spinal cord. As a result, the incidence of paralysis caused by polio declined (see Table 5.1). However, Salk's vaccine did not stop the spread of the virus, because it did not induce immunity in the intestine where the poliovirus initially reproduces. It was Albert Sabin who developed an oral vaccine, which induced immunity in the intestine, and which lessened the transmission of the virus.

By 1963, Sabin's oral vaccine had replaced Salk's injectable vaccine and by 1994, polio was eliminated from the Western Hemisphere. However, because the viruses in Sabin's vaccine were weakened, not killed, there were two kinds of unfortunate events. In some cases, the viruses could reproduce in the intestines, evolving due to new mutations into new strains that subsequently circulated and caused paralysis in unvaccinated people (cVDPV: circulating vaccine-derived poliovirus). In other, very rare, cases, they could cause poliomyelitis in the children who received the vaccine (VAPP: vaccine-associated paralytic polio). Importantly, the polio strains that have evolved from Sabin's vaccine nowadays cause more cases of polio in the world than the wild-type strains (see Figure 5.1).[21] In particular, since 2000, type 2 wild poliovirus has been causing 86% of cases of cVDPV. Because this type was eradicated in 1999, there is no need to protect children from it. Therefore, since 2016, there has been a change from a vaccine that contained weakened forms of all three poliovirus types to a form that does not contain the type 2 components in order to reduce the risk of cVDPV.[22]

But you may now be wondering: Should people be vaccinated against polio or not? To answer this question, it is necessary to consider the risks involved. And to understand risk, it is useful to consider another well-studied disease: cancer.

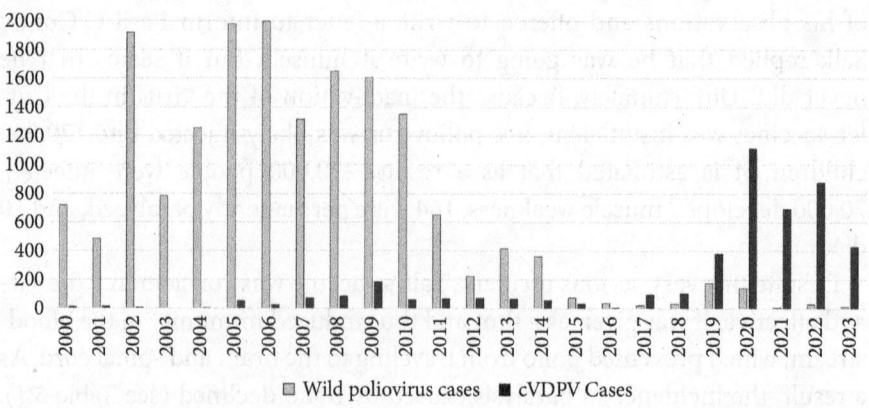

□ Wild poliovirus cases   ■ cVDPV Cases

Figure 5.1 Wild poliovirus (WPV) and circulated vaccine derived poliovirus (cVDPV) cases globally since 2000. It is evident that in recent years, more cases of polio are due to the vaccine than to the wild poliovirus (data from https://extranet.who.int/polis/public/CaseCount.aspx)

Uncertainty and risk in cancer

It feels nice to be certain about things and not to have to worry about what is to come. When we feel so, we are psychologically certain about things. Psychological certainty is about how strongly we believe something; we are psychologically certain when we are completely convinced that something is the case, beyond all doubt. But there is also epistemic certainty. We are epistemically certain of something when our evidence is so strong that it makes it *impossible* that we could be wrong. But much as we have the psychological need to feel certain, scientific knowledge does not require epistemic certainty; in fact, it is impossible to achieve epistemic certainty, because there will always be things that we do not or cannot know. This may not sound too good, but actually it is. In fact, uncertainty in science helps us to achieve the central epistemic aim of science: understanding. Because of epistemic uncertainty, scientists are continually searching for new evidence. If we were certain about something, we would stop researching it. Such evidence can provide us with even better support for the theories we accept, or it can lead us to new theories, the result being a deeper understanding of the natural world.[23]

This, of course, does not address our worries about the risks that may impact your life, and the lives of those around us. So what we need to be able to do is to understand and assess the risks relevant to a situation of interest. Psychologist Gerd Gigerenzer has suggested that uncertainty is a risk when it can be expressed as a number, such as a probability or a frequency based on empirical data.[24] To be able to quantify risk, the related alternatives, consequences, and probabilities must be known, as in the case of lotteries and other chance games. For instance, if there are 1,000 lottery tickets and there can be only one winner, then the probability of winning is 1 in 1,000. When we deal with alternatives, consequences, and probabilities that are unknown, we face uncertainty. In short, risk refers to known risks, whereas uncertainty refers to unknown risks.[25] When it comes to scientific knowledge, it is only risk about which we can do something, and this is where logic and statistical thinking can help.

There exist two types of risk: absolute and relative. Absolute risk is the probability for a person to have a particular outcome, for instance develop a disease, over a period of time. Relative risk is defined with some point of reference in mind, for instance by comparing the risk of an outcome between a group exposed to some factor of interest and a group not exposed to it. For example, if the absolute risk to develop a disease for a group of people exposed to a factor is p and the absolute risk to develop a disease for a group of people not exposed to the same factor is q, then the relative risk is p/q. Let us consider a fictitious example. Imagine that for a group of people not exposed to factor f related to a disease, the absolute risk is 0.4 (q). This indicates that the probability for this group to develop that disease is 40% or that 40 out of 100 people of that group will develop the disease over a certain period. Imagine now that in another group of people exposed to f, the absolute risk of developing the disease is 0.8 (p). This means that 80%, or 80 out of 100 people, of this group will develop the disease. Given these absolute risks, we can calculate that the relative risk is 0.80/0.40 = 2. A relative risk of 2 practically means that people exposed to factor f have twice the risk to develop the disease than those who are not exposed to factor f.

Let us consider an actual case. In 2015, the World Health Organization classified red meat (referring to all mammalian muscle meat, including beef, veal, pork, lamb, mutton, horse, and goat) as carcinogenic, that is, having the potential to cause cancer. Red meat was classified as a Group 2A carcinogen, meaning that the evidence for its connection to cancer was limited and was based on statistical associations. However, processed meat was classified as a Group 1 carcinogen, which means that there is strong and sufficient epidemiological evidence for the connection to cancer (by the way, tobacco smoke is also classified as a Group 1 carcinogen, but this does not mean that tobacco smoke and processed meat are equally dangerous; it only means that there is strong evidence for the connection to cancer). The key message was that: "An analysis of data from 10 studies estimated that every 50-gram portion of processed meat eaten daily increases the risk of colorectal cancer by about 18%."[26] This may sound like a lot, until one considers what it really means. The report itself stated that "A meta-analysis including data from 10 cohort studies reported a statistically significant dose–response association between consumption of red meat and/or processed meat and cancer of the colorectum. The relative risks of cancer of the colorectum were 1.17 (95% CI, 1.05–1.31) for an increase in consumption of red meat of 100 g/day and 1.18 (95% CI, 1.10–1.28) for an increase in consumption of processed meat of 50 g/day."[27]

It is important to note that the results of meta-analyses are among the most important ones to consider. A meta-analysis is, in simple terms, the analysis of data from various independent studies on the same subject, with the aim to determine overall trends. It is absolutely normal in science that some studies might find a weak association, others a strong association, and others no association at all between two variables. Such differences among studies are to be expected due to the different features of each study (for instance, different participants, different sample sizes, or different methods). However, once we look at all the available data together, we can get a useful overview of what is going on (see also Chapter 7).[28] The relative risk of 1.18, which can be interpreted as an increase in colorectal cancer for an increase in consumption of processed meat of 50 g/day, is more likely to be true when it is the result of a meta-analysis than when it comes from a single study. Let us see how this relative risk is calculated.

In a cautious and audience-sensitive analysis, statistician David Spiegelhalter has explained that whereas the increase of 18% in the risk for colorectal cancer may initially look like a lot, it may not actually be so. Here is what happens: The probability of developing colorectal cancer among those not eating processed meat was estimated to be 1 in 16, or about 6%; the probability of developing colorectal cancer among those eating 50 g of processed meat per day was estimated to be 1 in 14, or about 7%. This entails that out of 100 people not eating processed meat, 6 would develop colorectal cancer, whereas 94 would not. Similarly, out of 100 people eating 50 g of processed meat per day, 7 would develop colorectal cancer, whereas 93 would not. Therefore, the odds of cancer would be 6/94 for those not eating processed meat and 7/93 for those eating 50 g of processed meat per day. Based on this, it is possible to calculate the odds ratio: the odds that an outcome will occur given a particular exposure (7/93), compared to the odds of the outcome occurring in the absence of that exposure (6/94). If you do the math, $(7/93)/(6/94) = 1.18$.[29] The odds ratio is a measure of association between an exposure and an outcome, and in this case corresponds to a relative risk of 18%. However, it must be noted that the absolute risks for colorectal cancer are rather low, between 6% and 7%. Therefore, whereas the increase in (relative) risk sounds big, the actual (absolute) risks are relatively small.

So, how does this apply in the case of polio? Here are the absolute risks for the various cases:

- *Wild poliovirus*: Paralysis occurs in *1 in 200* infections (type II was eliminated in 1999 and type III in 2012, leaving only type I to deal with).
- *Vaccine-associated paralytic polio (VAPP)*: Paralysis occurs in approximately *1 in 2.7 million* (or 1 in 2,700,000) doses of the oral poliovirus vaccine (usually at the first dose).[30]
- *Circulating vaccine-derived paralytic polio (cVDPP)*: It is difficult to make estimates for this as vaccination against polio can decrease the

incidence of cVDPP, which also depends on the population density.[31] A recent estimation, as of November 2023, is that over the previous 10 years, there had been 800 cVDPP cases for 10 billion doses of oral polio vaccine given worldwide. If you do the math, this is *1 in 12.5 million* (or 1 in 12,500,000).[32]

We thus see that it is approximately between 13,500 and 62,500 times less likely to get paralytic polio from a vaccine than from a natural infection. At the same time, the wild virus may be less likely to encounter than the vaccine. Depending on the situation, there is a risk calculation to make.

Risks are also important when it comes to cancer diagnosis. One common cancer among women is breast cancer. According to the American Cancer Society, the lifetime risk (that is, the absolute risk until the age of 80) to develop breast cancer is 12.8% or approximately one in eight. This may seem like a lot, but it is important to note that it is the risk across a woman's lifetime. The chance that a woman will die from breast cancer has been estimated to be about 1 in 40 (about 2.5%, or 0.025).[33] These estimations entail that about 87% of women will not develop breast cancer, and about 98% of women will not die from it. But some women do develop breast cancer. The test usually done for this purpose is digital mammography, which is a low-dose X-ray scanning of breast tissue that produces a digital image of it. This is a useful preventive measure, which is nevertheless not 100% accurate. According to the American Cancer Society, about 12% of women undergoing mammography require follow-up imaging or biopsy, but eventually most of them (95%) do not actually have breast cancer.[34] Finally, mammography can also give a result that looks normal although there is breast cancer. Overall, it is estimated that mammograms miss about one in eight, or 12.5%, of breast cancers.[35]

Let us break this down to see what it entails. Among 1,000 women, 120 (12%) are initially diagnosed with cancer, whereas 880 are diagnosed as being cancer-free. The subsequent follow-up tests however show that among the 120 women initially diagnosed with breast cancer, only 6 actually have it (for whom the initial mammogram gave a true positive test), whereas the remaining 114 (95%) do not (so the initial mammogram gave a false positive test). At the same time, 12.5% of actual cancers are not diagnosed. This means that besides the six cases who actually had breast cancer, there was one more that was not diagnosed (false negative result). As Table 5.2 shows, 114 women would be alerted without having breast cancer, whereas 1 would be reassured although she did have the disease.

Table 5.2 Mammogram results and their accuracy in detecting breast cancer: A breakdown of true positives, false positives, false negatives, and true negatives in a sample population

		Breast cancer		
		YES	NO	
Mammogram result	Positive	**6** (True Positive)	**114** (False Positive)	**120** (Total Positive)
	Negative	**1** (False Negative)	**879** (True Negative)	**880** (Total Negative)
		7 (Total with disease)	993 (Total without disease)	

These potentially false results in breast cancer diagnosis with mammograms are reflected by two key features of diagnostic tests:

- *Sensitivity*: This is the ratio of the number of correctly diagnosed people as having the disease (true positive results) to the total number of people with the disease.
- *Specificity*: This is the ratio of the number of correctly diagnosed people as not having the disease (true negative results) to the total number of people without the disease.

Thus, based on the table above, we can calculate that the sensitivity of mammograms is 86% as 6 out of 7 women with cancer will be correctly identified as such, whereas the specificity of mammograms is 89% as 879 out of 993 women without cancer will be correctly identified as such. It is important to note that the number of women who would worry for nothing (114 in 1,000, or 11.4%) is much higher than the number of women who would miss the diagnosis of an actual disease (1 in 1,000, or 0.1%). A false alarm of this kind may be something that we may decide we have to live with, given that mammography does find the majority of breast cancers.

Although in most cases breast cancer is not an inherited disease, it has a genetic basis. There is a well-known relationship between particular versions of the genes *BRCA1* and *BRCA2* and breast cancer. But these relationships are probabilistic. According to the American Cancer Society, a woman with these versions of *BRCA1* or *BRCA2* has, on average, up to a 7 in 10 chance of developing breast cancer by age 80.[36] But this still means that there will be some women who will not develop breast cancer despite having these versions of *BRCA1* and *BRCA2*. Given how much these versions of *BRCA1* and *BRCA2*

Table 5.3 Breast cancer risk based on the presence of pathogenic *BRCA1&2* mutations: A comparison of disease incidence

Breast cancer

		YES	NO	
	Present	100	100	**200** (Total pathogenic *BRCA1&2* present)
pathogenic *BRCA1&2*	Absent	7000	92,800	**92,800** (Total pathogenic *BRCA1&2* absent)
		7100 (Total with disease)	**92,900** (Total without disease)	

genes increase the probability of breast cancer, it seems that testing for these genes in particular cases is worth the effort, as in the case of actress Angelina Jolie whose mother died at a relatively young age from the disease.[37] Let's see what this means in practice. In the USA, about 1 in every 500 women has a cancer-related mutation in either her *BRCA1* or *BRCA2* gene. Among these women, 1 in 2 will develop breast cancer by the time they turn 70 years old, compared to only 7 out of 100 women in the general United States population.[38] This means, if you do the math, that among 100,000 women in the USA, 200 are expected to have a cancer-related mutation in a *BRCA* gene, and 100 of these 200 will develop breast cancer. Among the women who do not have a cancer-related mutation in *BRCA* genes, 7,000 will develop breast cancer. So, among 100,000 women, there exist:

- 100 with a cancer-related mutation in the *BRCA* genes and breast cancer;
- 100 with a cancer-related mutation in the *BRCA* genes, but without breast cancer;
- 7,000 without a cancer-related mutation in *BRCA* genes, but with breast cancer;
- 92,800 without a cancer-related mutation in *BRCA* genes, and without breast cancer.

These calculations are presented in Table 5.3:

We thus see that whereas the absolute risk for a woman who carries a pathogenic *BRCA1&2* to develop breast cancer is 50% or 0.5 (100/200), this concerns only 0.2% of women in a population (200/100,000). For the remaining 99.8% of women in that population, the absolute risk to develop breast cancer is 7% or 0.07 (7000/100,000). This entails that the relative risk for women carrying pathogenic *BRCA1&2* to develop breast cancer is 0.5/0.07 = 7.14. This relative risk practically means that women carrying pathogenic *BRCA1&2* have seven times higher risk to develop breast cancer than those who do not carry pathogenic *BRCA1&2*.[39]

Once we understand the risks, I believe that we can make better decisions. It is quite rare for a woman to carry the pathogenic *BRCA1&2* versions, but if she does there are decisions she could make about potential prophylactic measures. At the same time, a woman who does not carry the pathogenic *BRCA1&2* is not relieved from the risk of breast cancer because most women who develop breast cancer do not carry the pathogenic *BRCA1&2* (this would be 98.6% of them or 7000/7100).

What is crucial to understand? (and also teach in schools!)

Psychologist Gerd Gigerenzer has identified a key problem in school education: "The problem is that our educational system has an amazing blind spot concerning risk literacy. We teach our children the mathematics of certainty—geometry and trigonometry—but not the mathematics of uncertainty, statistical thinking."[40] Yes, indeed! The consequences of this have been established by empirical research. For instance, Gigerenzer conducted a study asking 750 people in Amsterdam, Athens, Berlin, Milan, and New York what the statement that there is a "30% chance of rain tomorrow" means. Participants were asked to choose one of the following three options: (1) it will rain tomorrow in 30% of the region, (2) it will rain tomorrow for 30% of the time, and (3) it will rain on 30% of the days like tomorrow. The researchers found that only in New York did the majority of participants select option 3, which is the standard meteorological interpretation. In all the other cities, this option was judged as the least appropriate. The preferred interpretation in the European cities was that it will rain tomorrow "30% of the time," which was followed by "in 30% of the area."[41] This simple case clearly illustrates some of the problems with the interpretation of probabilities. A main issue, Gigerenzer noted, is that people do not understand probabilities because the reference class (percentage of what?) is not always clearly communicated to them. It is, therefore, necessary that understanding probability become a priority in education, as well as that scientists learn how to communicate their probabilistic findings to lay people in a way that the latter can make sense of them.

Let us consider a classic example: coin tossing. The probability that a single toss of a fair coin will result in heads is 0.5, or 50%, or one in two; the probability for tails is exactly the same. The probability that a fair die will result in a 6 is 0.16666, or 16.666%, or one in six; the probability for landing on a side with any other number (1–5) is exactly the same. This is a kind of probability where reference is made to objects or systems, the features and properties of which are well known. There is another kind of probability with reference to systems, the features and properties of which are not well known, or not known at all. In this case, we must make observations or conduct experiments, to see how often we get specific outcomes, that is, figure out their frequency. An example is when a drug is administered during a clinical trial to volunteer patients and healthy controls in order to see its effects. Finally, there is also the kind of probability, where the 30% probability of rain example falls. Unlike the cards, the weather is not a well-known system; unlike the clinical trial, meteorologists cannot conduct any kinds of experiments on days with identical conditions and then count the outcomes. The "30% chances of rain" is only an estimation based on previous knowledge and models.[42]

So far, so good, you may think. But these are not really the kinds of probabilities we worry about. Rather, we are usually concerned about dangerous situations that might cause death, such as disease or accidents. I argue that only when we compare risks to one another, can we make sense of them and decide what to worry or not worry about. For instance, we can make sense of the risk of developing breast cancer just discussed if we compare it to the lifetime odds of dying from various causes, estimated by the National Safety Council, a leading nonprofit safety advocate that focuses on eliminating the leading causes of preventable death (Table 5.4). It should be noted that no odds could be calculated for dying due to lightning, being a railway or airplane passenger, or exposure to radiation because the respective deaths were very few.

All this leads us to a very important question: Who decides what is a high or a low risk? That some causes of death are more likely than others is very helpful for putting risks in perspective. But it does not tell us on its own when measures should be taken and what kind of measures these should be. This is where values come into play. Scientists can establish up to a certain level whether the evidence for the various risks is sufficiently good, although uncertainties will always remain. However, they cannot alone determine the measures that ought to be taken, because these depend not only on evidence but also on social values and norms. Scientists can thus advise governments and the public, but they are not in any way politically authorized to make decisions about public policy. What risk is high or low, and therefore acceptable to live with or not, is something that cannot be decided by scientists alone but by democratic and publicly accountable officials or other means for decision-making.[43]

Table 5.4 Lifetime odds of death for selected causes, United States, 2021 (source: https://injuryfacts.nsc.org/all-injuries/preventable-death-overview/odds-of-dying/)

Cause of death	Lifetime risk of dying
Heart disease	1 in 6
COVID-19	1 in 10
Breast cancer	1 in 40
Drug poisoning	1 in 44
Opioid overdose (accidental)	1 in 58
Motor-vehicle crash	1 in 93
Fall	1 in 98
Gun assault	1 in 208
Pedestrian incident	1 in 485
Motorcyclist	1 in 747
Fall on and from stairs and steps	1 in 1,577
Uncontrolled fire in building or structure	1 in 1,602
Drowning and submersion while in or falling into natural water	1 in 2,042
Choking on food	1 in 2,659
Bicyclist	1 in 3,546
Exposure to excessive natural cold	1 in 3,699
Exposure to excessive natural heat	1 in 4,655
Drowning and submersion while in or falling into swimming pool	1 in 5,186
Accidental gun discharge	1 in 7,944
Dog attack	1 in 53,843
Hornet, wasp, and bee stings	1 in 54,516

The problem has been elegantly summarized by pediatrician and vaccine-expert Paul Offit, director of the Vaccine Education Center at the Children's Hospital of Philadelphia: "... a choice not to get a vaccine is not a risk-free choice. It's just a choice to take a different risk."[44] There is uncertainty involved in getting a vaccine, as in any other medical intervention. Of course, not all vaccines are the same. But in the case of polio, there is a real problem. On the one hand, there have been 99% fewer cases of wild poliovirus since 1988, with types II and III having been eradicated worldwide. On the other hand, those who still get vaccinated in the attempt to achieve the complete eradication of polio face the risk of developing the disease against which they were vaccinated because of the vaccine itself. This uncertainty can confuse people and make them distrust science. Therefore, correctly communicating the issue of vaccine safety to the public is very important so that people can weigh the risks of side effects against the probability of being infected with a deadly pathogen.

For instance, the vaccines RotaTeq (Merck) and Rotarix (GlaxoSmithKline) have been extensively used to prevent rotavirus, the most common cause of severe childhood diarrhea worldwide, which has occasionally been deadly. The result has been a significant decrease in hospitalizations and emergency room visits for rotavirus (e.g., by more than 80% among immunized children in the United States), as well as in deaths from diarrhea (e.g., by more than 40% in Mexico). Yet in some cases, a small but significant increase in the risk of intussusception (when part of the intestine folds into the section immediately ahead of it) has also been detected. In the United States, there is evidence that intussusception can occur as a result of vaccination with either of these two vaccines in approximately one to five cases per 100,000 infants. But when this risk is weighed against the hospitalizations, emergency room visits, and deaths from diarrhea, it is considered acceptable. For the 4.5 million babies born each year, vaccination against the rotavirus is estimated to prevent approximately 53,000 hospitalizations and 170,000 emergency room visits for diarrhea at the cost of causing 45–213 cases of intussusception.[45] The risk of this adverse effect makes it essential to pay attention to how things develop so that appropriate actions are taken in time, if necessary.

This entails that what scientists are to be trusted for, or not, is finding sufficient evidence to determine the risks, such as the risk of developing a disease or having side effects after vaccination. But the decision about whether a vaccine should be administered cannot, and should not, be their decision alone to make because it depends on the context and the respective values. For instance, parents in malaria-burdened regions in Malawi in Southeastern Africa were willing to have their children vaccinated even when there were still concerns about the safety of the Mosquirix vaccine.[46] The decision between taking the malaria risk or the vaccine risk depends on the social and cultural norms, and there can be no objective grounds for making such a decision. The parents in that case preferred to take the risk of dealing with the side effects of the vaccine rather than with malaria. They thus chose to have their children avoid a terrible disease they were well familiar with by taking the risk of having them experience the less known side effects of the vaccines. Scientists in this case could have advised them about both kinds of risks, but the final decision was theirs to make.

6

"The government feels their convulsions might have happened anyway. Dr. Murphy disagrees ..."

Conflicting "expert" claims

The *Vaccine Roulette*

> DPT. The initials stand for Diphtheria, Pertussis, Tetanus. Three diseases against which every child is vaccinated. For more than a year we have been investigating the "P", the Pertussis part of the vaccine. What we have found are serious questions about the safety and effectiveness of the shot. The overriding policy of the medical establishment has been to aggressively promote the use of the vaccine. But it has been anything but aggressive in dealing with the consequences. ... This is for sure, the Whooping Cough, or Pertussis vaccine, is the most unstable, least reliable vaccine we give our children.[1]

In 1982, a nationally aired news documentary called "DPT: Vaccine Roulette" presented the stories of children with seizures and brain damage. The documentary was written, produced, and presented by Lea Thompson, a consumer reporter for WRC-TV in Washington, DC. Starring with her were the children themselves and their parents, who shared emotional stories about how all their problems began as soon as their children received the diphtheria– tetanus–pertussis, or DTP, vaccine, while their doctors had reassured them that the vaccine was neither dangerous nor the cause of their children's problems. Thompson conducted interviews with various scientists, some of whom argued that the parents were right, whereas others tried to defend the use of the vaccine.

The documentary is a 50-minute montage of parents' and scientists' interviews, interspersed with children crying or behaving differently than usual. The parents describe how their lives changed since their child developed

Trusting Science. Kostas Kampourakis, Oxford University Press. © Oxford University Press (2025).
DOI: 10.1093/oso/9780197787106.003.0006

seizures, convulsions, or mental retardation. The emotions are strong. Along the way, there are scientists who explicitly question the safety of the DTP vaccine, whereas others argue that the available evidence is not sufficient for establishing the vaccine as the cause of all these children's troubles. Towards the end of the documentary, Lea Thompson was positive that even though most children have no problems because of the vaccine, "... some children have suffered learning disabilities and severe brain damage as a direct result of the shot." She blamed the US government for over-looking evidence related to this, and concluded that her investigation had provided evidence that many doctors and nurses were misinformed about just which children are at risk if they received the vaccine, that most doc-tors who saw adverse reactions did not report them, that the vaccine had been allowed on the market with little effectiveness or safety data, and that the tests for determining its effectiveness or safety were faulty. She con-cluded: "Our objective has been to provide information so there can be an informed discussion about Whooping Cough ... The dilemma for parents remains."[2]

The parents thus had a difficult decision to make: Should they vacci-nate their child according to the governmental recommendations, taking the risk that the documentary *Vaccine Roulette* supposedly revealed? Or should they simply ignore the government and avoid the risk of the side effects of the vaccine? Perhaps the government was not to be trusted? But what about the scientists who were supposed to know better than anyone else? Weren't they the ones whom parents ought to listen to? What were their views? The documentary ended with a concluding word by the scientists Thompson had interviewed (they appeared in this particular order, and their titles are those that Thompson used in the documentary):

Dr. Larry Baraff, UCLA, Whooping Cough Researcher: "I would certainly vaccinate my child ... yes."

Dr. Jerome Murphy, Pediatric Neurologist: "I would probably advise against it ... if the rest of the community were getting the Pertussis."

Dr. John Robbins, FDA, Bureau of Biologics: "Much more is to be gained by immunizing the children with our current vaccines with its limitations, than by allowing our children to be exposed to contracting Pertussis."

Dr. Robert Mendelsohn, Pediatrician, Author: "I feel that the vaccine should not be used, because the vaccine today represents a much greater threat than the Whooping Cough itself does."

Dr. Bobby Young, Microbiologist: "I recommended in writing to my daughter, so she could take this letter to her pediatrician (we don't normally communicate that formally), that my grandsons receive the "D" and the "T" component of DPT but not the pertussis component."

Dr. Edward Mortimer, American Academy of Pediatrics: "I believe it should be given to every child in the United States with the exception of very rare children in which there is a specific reason not to."

Dr. Gordon Stewart, British Epidemiologist: "I believe that the risk of damage from the vaccine is now greater than the risk of damage from the disease."

Dr. Alan Hinman, Centers for Disease Control: "I don't believe we have reached a stage in this country with pertussis where we have approached the stage where vaccination is more hazardous than the risk of disease."[3]

What were the parents in 1982 supposed to conclude from this ending? They saw and listened to eight "Dr.s," four of whom advised against it, whereas the other four suggested that it was worth receiving it. From a journalistic point of view, this is a sign of fairness: an equal number of experts supporting each side of the debate. At first sight, nobody could blame Thompson that she did not try enough to represent both views. Vaccine proponents came from the American Academy of Pediatrics, the Food and Drug Administration, the University of California Los Angeles, and the Centers for Disease Control. Nobody could complain that the official view had not been represented. In the same spirit, she gave the opportunity to other scientists to express their opposing views, such as Mendelsohn, a pediatrician who described himself as a "medical heretic." A microbiologist, an epidemiologist, and a pediatric neurologist had joined him. Those opposed to the DTP vaccines had equally been represented.

But this is misleading. What Thompson had probably intended was to create doubt about the official recommendations. We cannot know her motivation. Perhaps she was well intentioned and concerned about the children's health; perhaps there were other motivations behind the documentary. Whatever the reason, what she definitely managed to create was a portrayal of a divided scientific community. It would be a reasonable reaction of any concerned parent to question the official recommendations after watching the documentary. Why were so many scientists, presumably experts on the topic, arguing against the DTP vaccine, or at least the "P" portion of it? Why would a pediatrician, an epidemiologist, a pediatric neurologist, and a microbiologist express their concerns, if they did not have good reasons to do so?

Before we address these questions, let us consider the context of this story. The vaccine is called DTaP[4] for diphtheria (a disease that can result in difficulty in breathing, heart failure, paralysis, or death), tetanus (it causes painful stiffening of the muscles and can lead to various problems such as having trouble swallowing and breathing, or death), and pertussis (aP),[5] also known as "whooping cough" (it can cause uncontrollable, violent coughing that makes it hard to breathe, and can be extremely serious for babies and young children, causing pneumonia, convulsions, brain damage, or death).[6] As we saw, it was pertussis, a respiratory disease caused by the bacteria *Bordetella pertussis*, that concerned Thompson. The introduction of pertussis vaccines in the 1940s resulted in the annual cases of the disease falling from more than 100,000 to fewer than 10,000 by 1965. However, during the 1980s, pertussis cases began to increase gradually, with more than 48,000 cases reported in the USA by 2012, and the numbers declining ever since.[7] As is evident in Figure 6.1, *Vaccine Roulette* was followed by an increase in the number of pertussis cases.

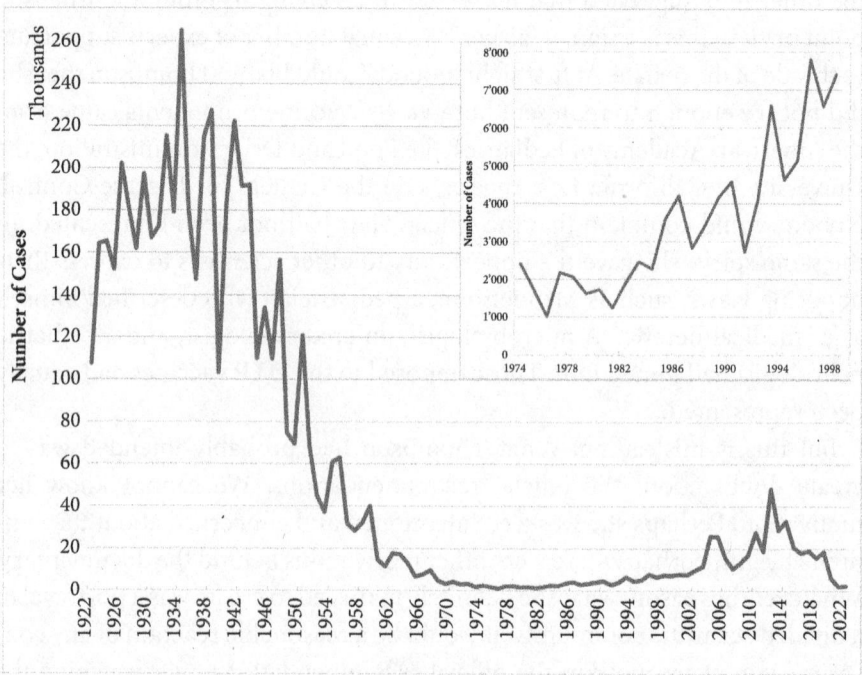

Figure 6.1 Number of pertussis cases reported to CDC from 1922 to 2022 in the USA. The introduction of the DTP vaccine in the 1940s brought about a significant decrease in the number of cases within a couple of decades. The number of cases does not seem to have increased very much after 1982 and the *Vaccine Roulette* with respect to what happened before the introduction of the vaccine. Still, compared to the years right before the documentary, there was an increase in the number of cases after the *Vaccine Roulette*. Data retrieved from https://www.cdc.gov/pertussis/surv-reporting/cases-by-year.html

A persistent controversy

Vaccine Roulette was not the beginning of the concerns regarding the vaccine against pertussis. It was in 1974 that a study in the UK, titled "Neurological complications of pertussis inoculation," raised questions about the safety of the DTP vaccine, as the abstract made clear: "Findings are presented in 36 children, seen in the past 11 years, who are believed to have suffered from neurological complications of pertussis inoculation (given as triple vaccine). The clustering of complications in the first 24 hours after inoculation suggests a causal rather than a coincidental relation." The neurological complications were caused by the DTP vaccine, the researchers claimed, because they appeared during the first day after vaccination. The data concerned 36 children between 1961 and 1972. Among these children, 2 died within six months of the onset of complications, 22 were mentally retarded and also suffered from epilepsy, 4 had mental subnormality without complications, 3 had epilepsy without mental retardation, 1 had persistent hemiparesis but developed normally otherwise, whereas 4 children recovered completely.[8] There had been previous case reports between 1948 and 1960 describing possible complications from the vaccine, which ranged from encephalopathy to permanent neurological injury and death. However, none of them had the impact that this 1974 study had.

TV documentaries and newspaper reports in the UK portrayed the dramatic stories of children who became retarded because of the vaccine. But there was more than the media. Parents in vaccine victim advocacy groups had an important role in sustaining the crisis. Perhaps most importantly, there was a strong divide within the medical community itself, most importantly between the Joint Committee on Vaccination and Immunization, which recommended the vaccine, and the physicians who actually administered it. A confrontational physician opposing the vaccine was Gordon Stewart, a medical professor at the University of Glasgow—who was also interviewed in *Vaccine Roulette*. In 1977, he published a study in which he noted: "Adverse reactions and neurotoxicity following vaccinations were studied in 160 cases. In 79, the relationship to pertussis vaccine was strong. In 14 of these cases, reaction was transient but characteristic of a syndrome of shock and cerebral disturbance, which, in the other 65 cases, was followed by convulsions, hyperkinesis, and severe mental defect."[9] The result of all the resulting mediatization was a decline in immunization rates against whooping cough from 80% in 1974 to 31% by 1978. This gave rise to a major epidemic; by 1979, 102,500 cases had been reported in the UK, more than any other period since the introduction of the vaccine (Figure 6.2), with 36 children dying. People seemed to have forgotten that the vaccine had been used in routine immunization since the 1950s and that during the 1940s, the disease had afflicted 60%–70% of children prior to completing school and had caused over 9,000 deaths.[10]

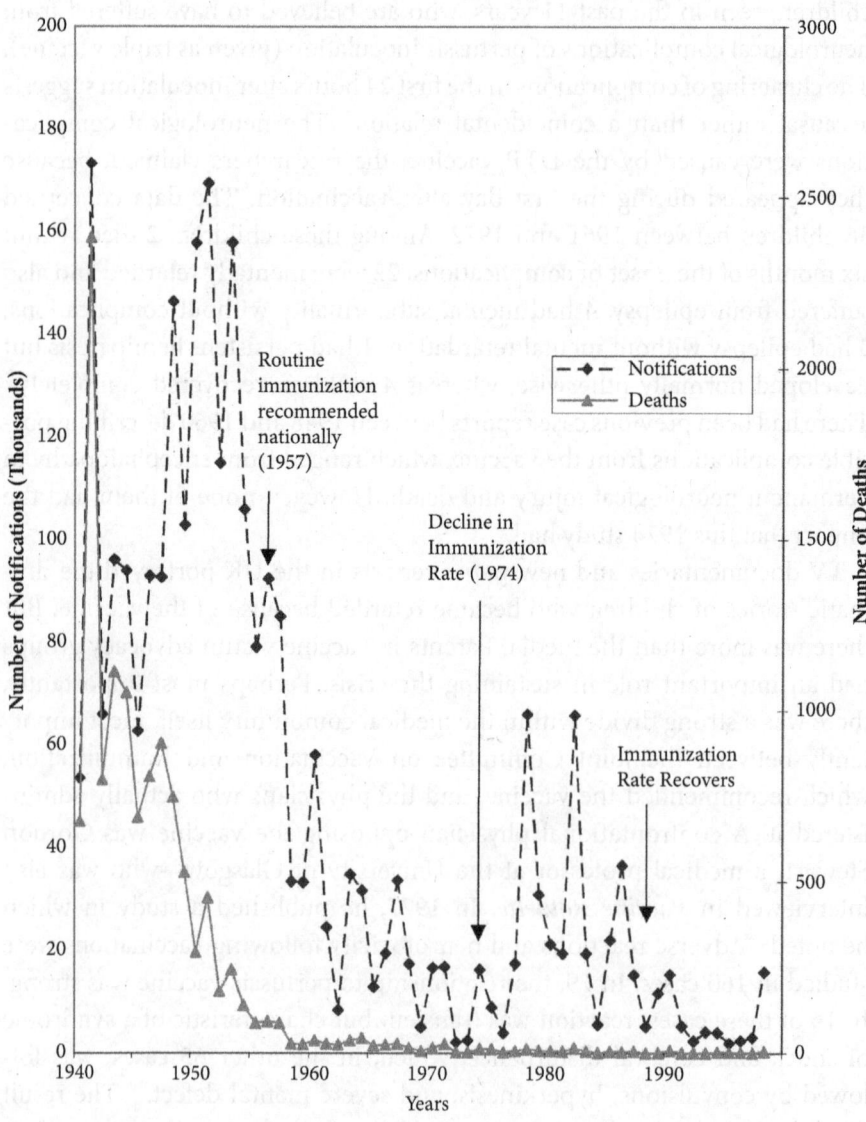

Figure 6.2 Whooping cough notifications and mortality—England and Wales, 1940–1998. Reproduced with permission from Baker, J. P. (2003). The pertussis vaccine controversy in Great Britain, 1974–1986. *Vaccine, 21*(25–26), 4003–4010.

Subsequent studies did not confirm the link between the vaccine and neurological damage. A 1981 study with 1,000 cases and 1,955 controls found a statistically significant association, but the authors noted that most children had recovered completely. They calculated that the risk of serious neurological disorders occurring within seven days after vaccination in previously normal children was 1 in 110,000 injections. However, they also noted that "The risk figures derived by the study must be set against the dangers of pertussis disease."[11] A couple of years later, a much larger study of 134,700 children who had received the DTP vaccine and 133,500 children who had received the DT vaccine (and therefore not the "pertussis" component) concluded that there was "No convincing evidence that DTP caused major neurological damage." The researchers noted that "If a striking syndrome of this sort exists, instances of it would probably have been discovered in the Northwest Thames Region where more than 400,000 injections of DTP were given in the seven years covered by the study."[12] More recent studies have confirmed this conclusion. For instance, a retrospective case-control study (see Box 5.1) based on records of 2,197,000 children between 1981 and 1995 attempted to estimate the relative risk of encephalopathy after vaccination with the DTP or the measles, mumps, and rubella (MMR) vaccines (see Chapter 7 for the latter) in the 90 days before the onset of the disease. A total of 452 encephalopathy cases were identified. The researchers' conclusion was: "In this study of more than 2 million children, DTP and MMR vaccines were not associated with an increased risk of encephalopathy after vaccination."[13] Instead of the vaccine being its cause, encephalopathy was considered to have a genetic basis as a particular molecular change was found in several children with alleged vaccine encephalopathy.[14]

You have probably noticed the common aspect in the 1974 UK and the 1982 USA stories: different experts disagreeing about the safety of the DTP vaccine. Of course, disagreement among experts does not occur in vaccinology only. Rather it may concern some fundamental questions, such as the cause of a disease. A striking example is the question about the origin of AIDS (which stands for acquired immune deficiency syndrome).

"Expert" disagreements about the cause of AIDS

If you lived through the 1980s, you would recall the two main fears of the era: the possibility of a nuclear war and AIDS. In the mid-1980s, several famous actors and other celebrities died of AIDS, and for some time it was mistakenly thought that the disease only concerned particular groups of people, such as homosexuals or intravenous drug users. It was soon shown that this was not the case. As with any new disease, it took some time for its cause to be established on solid evidence. Two key figures in the "discovery" of the cause of AIDS, which is the HIV (human immunodeficiency virus), were Robert Gallo and Luc Montagnier. This is a story of scientific controversy and debate that resulted in Montagnier, but not Gallo, receiving the 2008 Nobel Prize in Physiology or Medicine, along with Françoise Barré-Sinoussi "for their discovery of human immunodeficiency virus."[15] Barré-Sinoussi was Montagnier's collaborator and she was the one to find the evidence for the presence of a retrovirus in samples from AIDS patients, by identifying reverse transcriptase (an enzyme that produces DNA from RNA; this is considered to be old evolutionarily, hence the prefix "retro" in retroviruses. Reverse transcriptase produces from the RNA genome of the retrovirus DNA that in turn can integrate into the DNA of the affected cell). Montagnier himself has credited Gallo for the idea that a retrovirus was the cause of AIDS, after the scientific and legal controversies between them had faded. What mattered to Montagnier was that it took more than two years after AIDS was identified to show that HIV was its cause, and another two years for blood tests specific to it to become commercially available.[16] Gallo for his part acknowledged that a paper by Montagnier was unequivocally the first reporting a true isolation of HIV. He also noted that "The scientific achievements were overshadowed by a dispute between the United States and France over the patent rights to the blood test, and a temporary disagreement among the scientists."[17]

Gallo and Montagnier described how HIV was established as the cause of AIDS in a subsequent joint paper titled "The discovery of HIV as the cause of AIDS." Initially, several factors such as fungi, chemicals, and even an autoimmunity to white blood cells had been considered as possible causes of AIDS. However, scientists had noticed that the various manifestations of AIDS were all characterized by a decrease in the levels of a specific subgroup of T-cells that had the same antigen on their surface: CD4 (see Figure 2.4). It was therefore assumed that the agent responsible was one that affected T-cells with the CD4 antigen, such as the human T-cell leukemia virus or human T-lymphotropic virus (HTLV). Additional evidence supported this conclusion, as it was known from other studies that such viruses caused syndromes that were similar to AIDS. HTLV was transmitted through blood and sexual activity, as well as from mother to child, which was consistent with the ways of transmission of AIDS. Finally, the Centers for Disease Control and Prevention had reported cases of AIDS in people with hemophilia who had received only filtered clotting factors. That the agent could pass through filters was considered as evidence it was not larger than a virus. A key challenge with

AIDS was that it was characterized by clinical symptoms that appeared years after the infection. As a result, by the time the symptoms appeared, patients had also usually developed numerous other infections. Therefore, identifying the cause of AIDS presented a unique challenge. In the end, the hypothesis that AIDS was caused by a retrovirus was correct, but the more specific hypothesis that this was similar to HTLV was wrong. After 1984, it was HIV that was repeatedly isolated from patients with AIDS. A crucial step for this was the ability to grow HIV in T-cell culture lines, which in turn facilitated the development of a blood test for it. This test became available in blood-transfusion centers in 1985 and produced convincing evidence for the association between HIV infection and AIDS. It was thus widely accepted by the scientific community that HIV was the cause of AIDS.[18]

In 1988, however, Peter Duesberg, Professor of Molecular Biology at the University of California, Berkeley, published a one-page article in the prestigious journal *Science*, with the title "HIV is not the cause of AIDS." In that article, Duesberg argued that particular conditions, fundamental in order to establish a causal relation between a virus and a disease, were not met; for Duesberg, the idea that HIV was the cause of AIDS failed to meet some "... cardinal rules of virology."[19] His paper was accompanied by a point-by-point critical response by expert virologists, who noted that "Duesberg presents six (or nine) cardinal rules for viruses. Most are not relevant to the question of etiology and are misleading or wrong about viruses in general and HIV in particular."[20] This initiated discussions and debates on various issues, from whether Duesberg was right to whether he should be given the opportunity to express his view even if he was not right. The subsequent debates motivated the journal *Science* to conduct a three-month-long investigation in order to evaluate the claims made by Duesberg and his supporters. The final report was published in 1994, and its main conclusion was the following:

> This investigation reveals that although the Berkeley virologist raises provocative questions, few researchers find his basic contention that HIV is not the cause of AIDS persuasive. Mainstream AIDS researchers argue that Duesberg's arguments are constructed by selective reading of the scientific literature, dismissing evidence that contradicts his theses, requiring impossibly definitive proof, and dismissing outright studies marked by inconsequential weaknesses.[21]

In the meantime, Duesberg published more papers on this topic, two of which appeared in another very prestigious scientific journal, the *Proceedings of the National Academy of Sciences (PNAS)*. In the first of these papers, published in 1989, he made similar arguments with the 1988 *Science* paper, suggesting that "No such virus or microbe would require almost a decade to cause primary disease, nor could it cause the diverse collection of AIDS diseases. Neither would its host range be as selective as that of AIDS, nor could it survive if it were as inefficiently transmitted as AIDS." He thus proposed that AIDS was most likely due to a combination of pathogens, plus chronic drug use and malnutrition.[22] In the second paper, published in 1991, Duesberg further argued that the available epidemiological data did not support the idea that HIV was the cause of AIDS. He noted that HIV and AIDS were not always correlated; that in the USA, AIDS was almost restricted to males and that there was no exponential increase in the disease since the test was established, both of which are features of an infectious disease; and that HIV had long existed before.[23]

We therefore see here a scientist who went against the mainstream view, expressing his disagreement with the generally accepted view that HIV was the cause of AIDS.[24] Who was right? And how can we figure this out, and decide who to trust?

When experts disagree, who to trust?

Duesberg's views could not be easily dismissed: He had been elected as a member of the National Academy of Sciences in 1986,[25] and his views and arguments against the idea of HIV as the cause of AIDS had been published in two of the most prestigious scientific journals. When scientists have their work published in journals such as *Science* and *PNAS*, it is considered a huge achievement and a confirmation of its quality. Many questions thus arise: What should lay people have thought at the time? Who should they have believed, Duesberg or his critics? Can nonexperts assess the arguments and the evidence of expert scientists? Shouldn't work published in prestigious journals be considered as valid and reliable? Was Duesberg just a dissenter who aimed at bringing in new insights, contrary to the widely accepted view, as Galileo, Darwin, and Einstein before him had done? Indeed, this is perhaps what Duesberg thought of himself, as in his personal website, and in several of his papers, Duesberg begins with the following quotation, attributed to Einstein: "The important thing is to not stop questioning."

There are two things lay people can do in such cases. The first is to look for information related to the background of the publication of the various papers. For instance, a commentary published in *Science* soon after the publication of Duesberg's 1989 *PNAS* paper provided the background of its publication. All in all, it seems that Duesberg pushed hard to get his paper published, despite an initial rejection decision and subsequent negative reviews. The editor of the journal tried to be fair, and to consider Duesberg's paper without prejudice, even though his views had already been publicly criticized. Duesberg's paper was not considered original either, as it was seen as repeating the main points of the 1988 one in *Science*. It seems that in the end, the editor decided to give in to Duesberg's pressure, writing: "At this state of protracted discussion I shall not insist here—if you wish to make these unsupported, vague, and prejudicial statements in print, so be it. But I cannot see how this could be convincing to any scientifically trained reader."[26] Duesberg replied to this criticism by arguing that the review procedure followed was an appropriate one and that the editor's criticism was quoted out of context.[27] Notwithstanding Duesberg's claims, those scientists doing research on AIDS did not take him seriously, not only because he had never done any research on AIDS himself but also because all the available evidence contradicted his claims. As Anthony Fauci, whom we came to know during the COVID-19 pandemic, an AIDS researcher himself and director of the National Institute of Allergy and Infectious Diseases in 1988, put it at the time: "The evidence that HIV causes AIDS is so overwhelming that it almost doesn't deserve discussion anymore."[28] No AIDS researcher has disagreed ever since.

The second thing lay people can do is to consider what happened since then. If this was all just about theoretical arguments and inconsequential disagreements, there would be no big problem. One can believe and claim what one wants—in principle. But in Duesberg's case, his claims, perceived as coming from a credible scientist, may have caused the loss of numerous lives. In 2000, Thabo Mvuyelwa Mbeki, President of South Africa, appointed Duesberg to his "Presidential AIDS Advisory Panel." This of course boosted Duesberg's status and profile, but also had tragic consequences. When Mbeki succeeded Nelson Mandela as President of South Africa in 1999, the country already had almost one in five of its adults infected with HIV. Whereas about 300,000 people had already died from AIDS by that time, during Mbeki's presidency (1999–2008), many more people were added to the AIDS death toll. The reason for this was that Mbeki took seriously the dissident view that AIDS science was flawed and corrupted by the pharmaceutical industry. In one of his first speeches as President, Mbeki expressed his doubts about the safety of antiretroviral drugs, such as azidothymidine (AZT), and asked Health Minister Tshabalala-Msimang to investigate the issue. Two weeks later, she told parliament that although there were reports about the positive effects of AZT, there were scientists who were concerned about the toxicity of the drug that might outweigh the benefits of using it. At the same time, Tshabalala-Msimang rejected the reports by the Medicines Control Council (MCC) of her country that antiretroviral drugs were safe and effective, dismissing these instead as poison. Not only did she not support the MCC or the funding of antiretroviral drugs, but she also hired or supported people who advanced other remedies for AIDS. The outcome? It has been estimated that about 180,000 new HIV infections and 330,000 deaths could have been prevented during Mbeki's presidency.[29] It should also be noted that since 2010, the deaths have decreased by 70% and the number of newly infected people has decreased by 57%.[30]

Which raises the questions: Where did Mbeki go wrong? What should he have done, and whose advice should he have followed? Why was it a mistake to listen to the advice of Duesberg and other dissenters? Most importantly, what should lay people do when they come across disagreements and differing views among scientists, as in the DTP vaccine and AIDS stories presented so far?

What is crucial to understand? (and also teach in schools!)

Ideally, one should be able to discern real experts from supposed "experts," and follow the consensus view of the former. But how is a lay person supposed to know who is a real expert and who is not? What if it is the case that two "real" experts disagree? Would it be possible for a lay person to figure out that Duesberg was not an expert on AIDS as he had never done any research on this topic? Or evaluate the views of Mendelsohn with respect to the DTP vaccine? In any case, Duesberg was an expert on retroviruses and Mendelsohn was a pediatrician, so one cannot say that they had nothing to do with the topics they were commenting on. I suggest that lay people need not worry about this, as there is another way to approach the issue, one that is probably more effective. One should not listen to what individual scientists have to say on a topic, but rather figure out the consensus view of the scientific community. Figuring this out, though, is not simple or straightforward.

Nothing about the scientific community should be taken as self-evident: who belongs to it, who speaks for it, where it exists. It consists of individuals from various institutions, from universities to government bodies. Its members do not necessarily meet—this can happen occasionally at conferences—and it is impossible for any of its members to know, or to have met, every other member of it. Nor do all these members get along well with one another; it is possible to have different kinds of relations, ranging from close collaboration to ferocious competition. However, we commonly refer to the scientific community as one group because when it is necessary, as it has been in the cases of climate change and the COVID-19 pandemic, scientists have shown to be capable of consensus and concerted action. Based on her study of the scientific community during the last 350 years, intellectual historian Lorraine Daston has suggested some key factors that have led to successful scientific collaborations. The first is the establishment of standards to which everyone subsequently refers. Second, governance works best at the level of disciplines; interdisciplinary projects require careful coordination among the various disciplinary groups. Third, governments have consistently been the most reliable providers of sustained support for scientific projects. And finally, the bonds among the members of the scientific community are strengthened by personal interactions.[31] It is the second and the last point that are most relevant for our discussion here: why it matters to specify the discipline we are talking about, as well as to consider the interpersonal relations among scientists.

Although we colloquially refer to it, there is no single thing as "the" scientific community. Scientists have different kinds of disciplinary expertise and work on different and very specific topics. Therefore, when we refer to the scientific community with a specific topic in mind, we implicitly refer to the experts on that topic. For instance, when we refer to the views of the scientific community with respect to climate change, we do not refer to the same people we have in mind when we talk about the views of the scientific community with respect to vaccines. Therefore, what in fact exists are various scientific communities to which we refer when we use the term "scientific community." These communities may or may not overlap, depending on the topic, but any individual scientist can belong to only a few of them. The reason for this is that the area of expertise of each scientist has pragmatic limits, as there is only a small number of topics that one can learn and understand. Therefore, any scientist can only be an expert on very few topics. For instance, it is very unlikely for an expert on climate change to claim expertise on vaccines and vice versa, unless it is someone working on an area common to both, if that exists (mathematical modeling perhaps?). Furthermore, even within such communities, there always exist scientists with different areas of expertise working on the same topic from different perspectives.

In addition, even though it is practically impossible for any member of a scientific community to personally know every other member, it is usually the case that everyone knows the top experts in the field as well as that they themselves know one another. Whether they get along together well, or not, scientists usually know who is experienced, who has done research on specific topics, and overall whether one knows what one is talking about. It is this personal knowledge and interaction among scientists that provides the basis for the exchanges among them that may pave the way for reaching a consensus view. For instance, Montagnier and Gallo knew, and respected, each other and thus, despite their disagreements during the 1980s, they managed to have a rapprochement and publish a series of papers in 2002 and 2003, in which they expressed their appreciation of each other's work. Gallo also knew Duesberg well. When Gallo introduced Duesberg at a 1984 meeting of the National Cancer Institute in Maryland, USA, he began by referring to him as a friend and described him as "… a man of extraordinary energy, unusual honesty, enormous sense of humour, and a rare critical sense. This critical sense often makes us look twice, then a third time, at a conclusion many of us believed to be foregone."[32] But in later years, Gallo rejected Duesberg's questioning of HIV as the cause of AIDS.[33]

By being experts on a particular topic and by knowing "who is who" in their community, scientists can reach a consensus view. This consensus view can in turn be considered as trustworthy thanks to the internal procedures of critical interrogation within the community. For instance, peer review, properly implemented, offers a mechanism by which biases, errors, inconsistencies, and incompleteness can be identified and corrected or left out. The better scientists follow these procedures, the less problematic the conclusions of the scientific community will be. This in turn provides the basis for trust in the consensus view of the scientific community, but not in those of individuals. Only the scientific community and its internal procedures can guarantee the best understanding. But even in this case, the trust that lay people ought to have should be informed, not blind. They should demand experts to explain to them as much and as well as possible how and why they have reached their consensus view. And lay people ought to trust experts when they see that they diligently apply the various procedures of self-correction in order to achieve the best understanding possible.[34]

Whatever any individual scientist claims, whatever their status, cannot and should not matter more than the consensus view of the respective community of experts, which usually represents the majority. This is not an issue of democracy, though, but of carefully considering and interpreting the available data. There have been several "dissenters" in the history of science, that is, scientists who departed from and questioned the mainstream view of their time, and who were eventually shown to be right. This is why a dissenter's view should be carefully considered. If the scientific community had simply dismissed the claims of Duesberg, this would not be sufficient to trust them. But when the community has engaged with Duesberg's arguments, has considered them in detail, and has decided based on the available data that they are wrong, then there is really no reason any more for lay people to listen to Duesberg. Science is not done by individuals but by communities, and it is the consensus view of the community of expert scientists that we should trust.

Perhaps the clearest example of scientific consensus is that about human-caused climate change. Historian of science Naomi Oreskes has analyzed the abstracts of 928 peer-reviewed scientific journal articles published between 1993 and 2003, which were listed in the Institute for Scientific Information database with the keyword "climate change." She found that none of the papers disagreed with the consensus position that climate change is occurring and that it is being strongly influenced by human actions.[35] Furthermore, John Cook, Professor at George Mason University's Center for Climate Change Communication, and his colleagues conducted a similar analysis in which they examined the abstracts of 11,944 peer-reviewed scientific journal articles published between 1991 and 2011 with topics on "global climate change" or "global warming." They found that when it comes to the consensus position that humans are causing global warming and climate change, it was endorsed in 97.1% of the analyzed articles whereas it was rejected in only 0.7% of them.[36] In more recent years, a review of 11,602 peer-reviewed articles on "climate change" and "global warming" has concluded that the scientific consensus on human-caused global warming has grown to 100%.[37]

I am inclined to think that it is not likely to find such an absolute agreement on most scientific topics. But I do not believe that this is necessary. Whatever view you find that the vast majority of expert scientists have, this is the one to follow. And you do not need a review of the literature or a survey for that. The various academies of science or scientific associations usually make formal statements that represent the consensus view on a topic. But what should we do when such consensus is not clearly expressed, or does not even exist?

Philosopher Elizabeth Anderson has suggested that even though most lay people cannot judge the merits of most scientific claims, they can decide what to believe by judging whom to believe. For these judgments to be possible, it is necessary to have criteria of trustworthiness and consensus that lay people can easily apply, using information they can easily access. These judgments in turn depend on the assessment of:

1. *expertise*, that is, judge whether scientists have access to the evidence and the skills to evaluate it;
2. *honesty*, that is, be able to judge whether scientists tend to honestly communicate what they believe, and refrain from misleading their audience;
3. *epistemic responsibility*, that is, judge whether scientists consider evidence and arguments that counter their own beliefs.[38]

In order to be able to assess expertise, honesty, and epistemic responsibility, Anderson has proposed particular criteria that are summarized in Table 6.1. The higher a scientist is found in the hierarchy of scientific expertise for a particular domain, and the fewer factors casting doubt on one's honesty or indicating epistemic irresponsibility we find, the more trustworthy that scientist should be considered. Next time you come across two opposing views, take this list out and start noting what you see. You will be surprised by how helpful it will be.

Hoping that I have made a compelling argument about how important it is for everyone to discern and honor the consensus within the respective specific scientific community, we need to keep in mind that this consensus might take some time to develop. Perhaps today we are convinced that a particular vaccine is safe, but the key question is what people can know when a controversy erupts. This is why it is necessary for the respective scientific community either to intensively publicize whatever their consensus view is or explain why they have not yet arrived at one. In either case, scientists have a social responsibility to take positions and inform lay people about the current status of their understanding. If they do not, then someone else might do so in their place and exploit the situation.

Bioethicist Bernard Rollin has drawn on what he has called the "Gresham's law for ethics," inspired by the monetary principle of Thomas Gresham, to explain the dynamics in public discussions on science and technology. Gresham's law states that less trustworthy currency drives out more trustworthy currency in an economy. For example, if people can use both gold coins and paper money to buy goods, they are likely to view gold coins as more valuable because they have intrinsic value as gold in addition to their monetary value. In contrast, paper money only holds monetary value. As a result, people will prefer to save their gold coins and spend their paper money, leading to the circulation of gold coins diminishing while paper money becomes more prevalent. In a similar way, Rollin has argued, ill-informed and biased ethical

Table 6.1 Criteria that lay people can use to assess expertise, honesty, and epistemic responsibility (based on Anderson, E. (2011). Democracy, public policy, and lay assessments of scientific testimony. *Episteme, 8*(2), 144–164, 145–148)

Criteria for judging scientific expertise	Criteria for judging honesty	Criteria for judging epistemic responsibility
Where in the hierarchy of scientific expertise for a particular domain is one found?	*Does any of the factors that cast doubt on one's honesty, and may thus discredit one's claims, exist?*	*Does any of the factors that indicate an evasion of accountability, and therefore epistemic irresponsibility, exist?*
• Scientists who are leaders in the domain and have been officially recognized as such • Scientists whose current research is widely recognized and cited by other experts in the domain • Scientists who are research-active in the field • PhD scientists trained in the domain • PhD scientists outside the domain, but with relevant expertise • PhD scientists outside the domain • People with a science degree far removed from the domain • Lay people	• Conflicts of interest, such as receiving funds from agents who have interests in getting people to believe a particular claim • Evidence of previous scientific dishonesty, such as plagiarism, faking experiments, or data, and repeatedly citing research that does not support one's claims • Evidence of misleading statements, such as cherry-picking data, or taking quotations out of context • Persistently misrepresenting the arguments of opponents or making false accusations of dishonesty against them	• Evasion of peer-review: refusing to share data or to reveal methods and procedures to allow replication; failing to submit research to peer-reviewed journals; publicizing one's ideas before sharing it with experts • Dialogic irrationality: continuing to repeat refuted theories, without responding to the refutations • Advancing refuted theories in domains other than the one under investigation—for example, that HIV does not cause AIDS • Voluntarily associating with supporters of refuted theories—for example, publishing their work, or placing one's own work for publication in their venues

arguments in debates on new technologies can overshadow well-reasoned and informed ones. Early discussions about new technologies lack sufficient experience to foresee all potential ethical and societal issues, making it crucial for experts to provide balanced, well-informed perspectives. If they don't, the debate may be dominated by those with extreme views or ideological biases. Rollin emphasized the professional duty of experts to educate the public and guide discussions, ensuring that debates on new technologies are grounded in sound reasoning.[39]

When it comes to scientific knowledge, it is the real experts who understand any given situation better than anyone else, and it is only they who can guide us to make informed and appropriate decisions. Therefore, we need to be able to identify who these experts are and devote the time to doing so before making any decisions about socio-scientific issues.

7

"The medical profession was wrong, in some cases shamefully so"

Flawed science and unreliable review procedures

The "MMR scare"

> The approach of the clinical scientists should reflect the first and most important lesson learnt as a medical student—to listen to the patient or the patient's parent, and they will tell you the answer. ... The parents were right. They have helped us to identify a new inflammatory bowel disease that seems to be associated with their child's developmental disorder. This is a lesson in humility that, as doctors, we ignore at our peril. In many cases, the parents associated onset of behavioural symptoms in their child with MMR vaccine. Were we to ignore this because it challenged the public health dogma on MMR vaccine safety? ... Assumptions of vaccine safety, based upon inadequate safety trials and dogma contribute largely to confusion and public loss of confidence in vaccination.[1]

This is Andrew Wakefield, a surgeon working at the time at the Royal Free Hospital School of Medicine in London.[2] On February 26, 1998, he presented during a press release the results of a study that implied a connection between the measles, mumps, and rubella (MMR) vaccine, gut disorders, and autism in children. MMR is a triple vaccine against measles, mumps and rubella, which are highly contagious viral diseases for which there is no specific treatment. This is why vaccination is the only effective means of prevention, and it is recommended that all children be vaccinated with two doses of the MMR vaccine between 12 and 24 months of age. Two days later, a paper on the topic was published in the prestigious medical journal *The Lancet*, with Wakefield as its first author. In the summary of the paper, it was mentioned that 12 children between 3 and 10 years old were referred to the hospital "with a history of normal development followed by loss of acquired skills, including language,

Trusting Science. Kostas Kampourakis, Oxford University Press. © Oxford University Press (2025).
DOI: 10.1093/oso/9780197787106.003.0007

together with diarrhea and abdominal pain." It was noted that the onset of these symptoms "... was associated, by the parents, with measles, mumps, and rubella vaccination in eight of the 12 children ..." The interpretation given in the summary was this: "We identified associated gastrointestinal disease and developmental regression in a group of previously normal children, which was generally associated in time with possible environmental triggers" (the latter most likely being an implicit reference to the MMR vaccine).[3] The message at the beginning of the paper seems clear: The MMR vaccine might be causing gut and behavioral problems.

However, and interestingly, the conclusion that Wakefield and his coauthors reached towards the end of the paper was different. "We did not prove an association between measles, mumps, and rubella vaccine and the syndrome described. Virological studies are underway that may help to resolve this issue. If there is a causal link between measles, mumps, and rubella vaccine and this syndrome, a rising incidence might be anticipated after the introduction of this vaccine in the UK in 1988. Published evidence is inadequate to show whether there is a change in incidence or a link with measles, mumps, and rubella vaccine."[4] So we see not only that Wakefield and his colleagues were explicit that the evidence they presented was not adequate to "prove" an association between the MMR vaccine and the observed symptoms; they also made the reasonable assumption that if this had been the case, there should have been a rise in the incidence (that is, the number of new cases of disease in a population) since 1988 when the vaccine was introduced in the UK.

In the very same volume of *The Lancet*, there was a commentary written by Robert Chen and Frank DeStefano, who both worked in the Vaccine Safety and Development Activity National Immunization Program, of the Centers for Disease Control and Prevention. They noted that since the mid-1960s, hundreds of millions of people all over the world had received a vaccine containing the measles virus (an RNA virus) without developing either gut or behavioral troubles. In contrast to Wakefield and his colleagues, who considered the evidence for this in the UK insufficient to arrive at a conclusion, Chen and DeStefano noted that the available evidence contradicted the claim made by Wakefield and his colleagues. If the MMR vaccine did indeed cause the syndrome observed, they added, this should have been something happening "extremely rarely." They also pointed out that as the first dose of the MMR vaccine was given to about 600,000 children every year in the UK, usually during their second year of life when autism also first becomes manifest, it should not be surprising that some cases of autism appeared after MMR vaccination.[5] Finally, they referred to another study in the same volume of *The Lancet* that did not find any trace of the measles virus in patients with the same symptoms as those that Wakefield and his colleagues had reported.[6] Even more astonishingly, a study with Wakefield as its last author that was published in December of the same year reported similar results.[7]

These are the criticisms that brought about Wakefield's response, quoted at the beginning of the present chapter. But instead of relying on scientific evidence, Wakefield relied on emotion to make his case. He painted a picture of a relation of inequity and snobbism that existed between the medical establishment on the one hand and the parents on the other. In this picture, the parents are helpless against the Goliath that the dogmatic medical establishment represents. This is where Wakefield gets into the picture. He enters the scene from within the medical establishment as the maverick scientist who challenges the prevailing dogma, as another David who comes to defeat the Goliath of the medical establishment. There was no scientific argument to be taken seriously, mostly because as Chen and DeStefano—and even Wakefield himself—suggested there was no evidence after millions of vaccinations that supported the connection between vaccines and autism. So, all that Wakefield could do was to rely on emotion, perhaps in an attempt to get the support of the parents of children with autism.

Initially not much happened following the publication of Wakefield's paper in 1998. But in 2001, the story began to gain publicity. As physician and science writer Ben Goldacre has explained, Wakefield continued to publish on the topic. It is also possible that particular newspapers and people were interested in attacking the government and the healthcare system. This led to an increase in newspaper articles about the MMR vaccine. The ensuing discussion was not at all about the scientific evidence. But what made the topic explode in December 2001 was the question of whether Leo Blair, the baby son of the then prime minister of the UK, had received the MMR vaccine or not. When asked, the Blairs refused to answer, arguing that their son's healthcare was a private matter.[8] This resulted in a fall in MMR vaccine coverage in the UK, but without a big rise in the number of cases and deaths. Goldacre insisted that even though Wakefield was to blame for what happened, it was the media that created what Goldacre described as the "MMR hoax" and that maintained it diligently for 10 whole years: "It was the perfect story, with a single charismatic maverick fighting against the system, a Galileo-like figure; there were elements of risk, of awful personal tragedy, and of course, the question of blame. Whose fault was autism?"[9]

Public health scholar Tammy Boyce has studied closely the coverage of the MMR–Wakefield story in the UK press. As is evident in the left part of Figure 7.1, there was an association between the increase in the coverage of the MMR story in the UK newspapers and the decline in MMR vaccine coverage. In contrast, this does not seem to have been the case in the USA, where the MMR story attracted a lot less attention. The coverage of the MMR story in the UK media, and the introduction of the idea that the triple MMR vaccine might not be safe, led parents to question the safety of a vaccine they had previously not worried about. According to Boyce, there were multiple reasons why the media raised concerns about the MMR vaccine and why this had the negative influence that it had. To begin with, earlier socio-scientific controversies about the "mad cow disease" (bovine spongiform encephalopathy) and genetically modified organisms had set the frame for the MMR story to appear as a controversy. Furthermore, these earlier controversies had made the Department of Health seem untrustworthy. Most importantly, the story quickly became political rather than scientific. It became mostly about the politics of healthcare provision rather than about science. Therefore, the focus was not so much on the scientific evidence about the MMR vaccine, but about parental choices with respect to vaccination. This made the involvement of scientists more difficult than what it already was. In addition, as their views were aligned with that of the government, they were not considered as newsworthy by the numerous journalists who reported on this story even though most of them had no

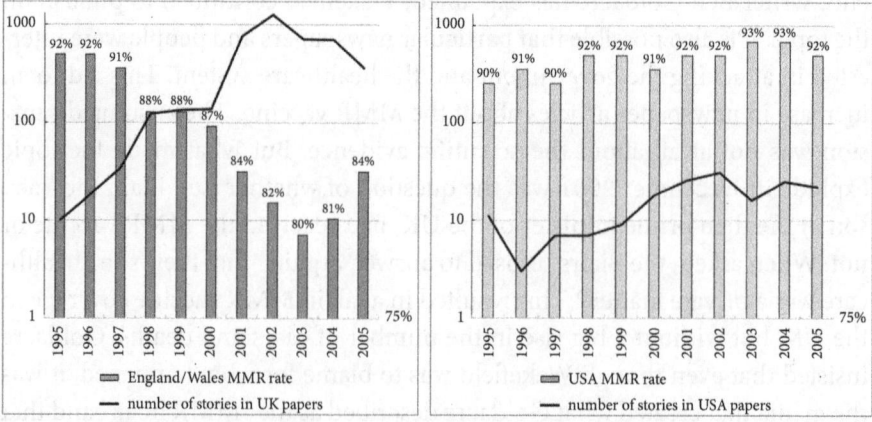

Figure 7.1 Number of articles containing "MMR" in UK/ USA newspapers, along with the respective MMR vaccine coverage in these countries. *Left*: An association seems to have existed between the increase in the coverage of the MMR story in the UK newspapers and the decline in MMR vaccine coverage. The more the number of stories increased, the more the vaccination levels fell. *Right*: No such association was observed in the USA, where the number of articles containing "MMR" were a lot less, and the MMR vaccine coverage less variable (data from Table 1.1 of Boyce, T. (2007). *Health, Risk and News: The MMR Vaccine and the Media*. New York: Peter Lang.

background in science. Presenting the parents' concerns about the vaccine's safety was also easier and catchier than explaining the official position about the lack of evidence for the connection between the MMR vaccine and autism (see Chapter 8 on this issue), or other health problems. Finally, journalists often equated parents' accusations or narratives with scientific arguments. As Boyce noted, "... if scientists leave ethical or controversial debates only to those without scientific expertise and knowledge, there will be an absence of science in these decisions, thus affecting the tone of the debate" (a point already made with Gresham's law in the previous chapter).[10]

Notwithstanding the problems caused by journalists and the media, it was the detailed investigation by journalist Brian Deer that revealed in 2004 all kinds of problems with the 1998 Wakefield paper. Whereas children had been presented in the *Lancet* article as being autistic, the respective medical records did not show that they had been diagnosed with autism. Also, whereas children had been presented in the article as having the first symptoms within days of the MMR vaccine, the medical records did not mention anything like this for months after vaccination. Furthermore, two years before the publication of the *Lancet* paper, Wakefield had been hired by the attorney Richard Barr who wanted to raise a speculative class-action lawsuit against the pharmaceutical companies that produced the MMR vaccine. In 1996, Wakefield was awarded an initial sum of £55,000, and charged his consultations at the extraordinary rate of £150 per hour (billed through his wife's business), his earnings eventually totaling £435,643. Finally, in June 1997, nearly nine months before the press conference at which Wakefield had proposed a single vaccine, he had filed a patent on products, including his own "safer" measles vaccine.

Deer's revelations resulted in Wakefield losing his medical license in 2010, having already been asked in 2001 to resign from the Royal Free Hospital. In 2004, he moved to the USA to become the director of research for the International Child Development Resource Center in Melbourne, Florida. But a year later, he moved again to work for the Thoughtful House Center for Children in Austin, Texas. However, he also resigned from that in 2010 due to the disciplinary hearings of the General Medical Council in London, which resulted in his name being erased from the medical register. The procedure of the General Medical Council, which lasted from July 2007 to May 2010, arrived at the conclusion that Wakefield's conduct was irresponsible and dishonest. Deer's revelations also resulted in Wakefield's *Lancet* paper being retracted in the same year—a whole 12 years after its publication.

As a result of the 2004 revelations made by Deer, many of Wakefield's coauthors retracted their support of the paper's main conclusion. Wakefield's own reaction in 2004 was, not surprisingly, along the same lines as his 1998 defense of his paper:

> Let us be clear that parents reported gastrointestinal symptoms in their children that many medical professionals denied and refused to investigate. Some parents were referred to social services and false claims of Munchausen's syndrome by proxy were levied. The parents were right; their children have an inflammatory intestinal disease. The medical profession was wrong, in some cases shamefully so. In light of this lesson it is imperative that rather than relying on endless reviews of epidemiological data which fail to even address the original hypothesis, parental claims should be taken seriously and their children should be investigated on an individual basis.[11]

Peer review, and why it matters

The editor-in-chief of *The Lancet*, Richard Horton, had this to say after the retraction of the 1998 paper by Wakefield:

> Peer review is the best system we have got for checking accuracy and acceptability of work, but unless we went into the lab or examined every case record, we can't ever finally rule out some element of misconduct. The entire system depends upon trust. Most of the time we think it works well, but there will be a few instances—and when they happen they are huge instances—where the whole thing falls apart.[12]

The process of peer review is key to the success of science, and it is also an important feature of its being a social process. No scientific work is considered credible if it has not undergone peer review, because this process is expected to filter out the good work from the not so good one. This is the stage where it is decided what is published and how each scientific field proceeds. However, subjectivity and biases are to be found in this process because it is done by human beings. These do not have to be deliberate; what matters is that they are there and we are aware of them.

In the same interview, however, Horton also said: "It was utterly clear, without any ambiguity at all, that the statements in the paper were utterly false ... I feel I was deceived."[13] But had Horton taken any measures during the peer-review process to avoid these problems? In his personal, book-length account of the Wakefield affair, Horton noted that he and the other editors of *The Lancet* had "... tried to highlight the preliminary nature of these findings by publishing the [Wakefield's] article as an 'Early Report.'" They also tried, he added, to remind readers of the vast benefits of measles vaccination at the global level. Horton also noted that "Two respected vaccine experts, Robert T. Chen and Frank DeStefano, invited readers to examine the Wakefield data with 'an open mind.' But they obviously doubted the possibility of the alleged association,"[14] as we also saw in the previous section. This raises many questions: Did Horton invite Chen and DeStefano to write the commentary? This is most likely, as it appears in the same volume with Wakefield's article, so they must have known about it before its publication. Was any of them involved as a reviewer of the manuscript? If yes, it is unlikely that they would have said anything different in the review than what they wrote in the commentary.

Horton has elsewhere written that "In the case of Wakefield's paper, four reviewers made favorable comments, although questions about the methods used and interpretations drawn were raised—and justifiable concern was expressed about the impact of the findings on the public's confidence in a very important vaccine."[15] Chen and DeStefano had written in their commentary that "... vaccine-safety concerns such as that reported by Wakefield and colleagues may snowball into societal tragedies when the media and the public confuse association with causality and shun immunisation."[16] They were right, of course. As Horton himself admitted in 2004, "The Wakefield paper was the beginning of a sequence of events that saw their concern become a reality."[17] The year before (and before the revelations about Wakefield and his coauthors' retraction of support of the paper's main conclusion), Horton had already taken responsibility for what had happened since it was its publication in *The Lancet* that gave that paper more credibility than it had deserved. "There would have been no press conference or press release, video or otherwise, if *The Lancet* had not published the study in the first place." Horton admitted that he had not expected what followed, that he had been "terribly and, ... embarrassingly naïve," that he "... failed to do enough to manage the media reaction to that work," and that until that time he "had not seen this media management role as one for a scientific medical journal editor," whereas he thereafter considered it as one of his main responsibilities. But, after having written all that, Horton added: "But I do not regret publishing the original Wakefield paper. Progress in medicine depends on the free expression of new ideas. ... Verification is the right test of new thinking, not censorship."[18]

The process of peer review in science is perhaps the most important one for ensuring the quality of the findings published. In principle, scientists should publish their findings in order to share them first with the scientific community, and then—via journalists and other science communicators—with lay people, who often, as taxpayers, indirectly fund scientific research.

Before this happens, their peers are invited to read and comment on their work, which is presented in the form of a scientific paper. This usually happens in a double-blinded manner, that is, during the review process the identities of neither the authors nor the reviewers are disclosed. Only the journal editors who process the submitted papers and assign them to reviewers know who is who (however, in recent years, this has begun to change in various ways). The peer-review process can be one of the most subjective endeavors in the scholarly world, because reviewers have their own conceptualizations, views, experiences, and biases, which can collectively impact the stance taken towards a manuscript. This is where a professional attitude is crucial, in order for reviewers to produce as objective an evaluation as possible of the work under consideration.

But peer review is not a process without problems. The first problem emerges from the reasons that scientists publish. As the number of publications one has, as well as the prestige of the journals one has published in, matter for getting promotions and grants, many scientists have developed a "publish-or-perish" mentality. For instance, in the life sciences, if you publish many articles in journals such as *Cell, Nature, Science, The Lancet*, and *The New England Journal of Medicine*, you are considered the top in your field, and this may get you better job offers or funding. But if you do not, you may lose in the competition with others. This is what unfortunately has made several scientists commit fraud. This can take many forms from manipulating data to editing photographs and other visuals in order to get the desired results. Unfortunately, these fraudulent actions are not always easy for the reviewers to spot. But when fraud is uncovered, a common outcome is retraction. This is the case for even highly cited papers, published in prestigious journals (Wakefield's 1998 *The Lancet* paper is the fourth most cited paper among the retracted ones).[19]

According to Ivan Oransky and Adam Marcus, the creators of Retraction Watch, a blog launched in 2010 that reports on retractions of scientific papers, retractions are a window into the scientific process. One of the most important features of science is its self-correcting nature; seeking more or better data, conducting better analyses (I will come back to this in a moment). "But when a retraction is necessary, how long does that self-correction take? The Wakefield retraction, for example, was issued 12 years after the original study, and six years after serious questions had been raised publicly by journalist Brian Deer."[20] A new record was made in 2023 with more than 10,000 research articles being retracted. Among those, more than 8,000 retractions concerned articles published by journals of a single publisher (Hindawi, a subsidiary of Wiley) due to "concerns that the peer review process has been compromised" and "systematic manipulation of the publication and peer-review process."[21]

But do most scientists commit fraud? In 2009, Daniele Fanelli published a meta-analysis of 18 studies that had directly asked scientists whether they had committed scientific misconduct or knew of a colleague who had. Scientific misconduct consists principally of (1) fabrication, such as inventing data; (2) falsification, which is the willful distortion of data or results; or (3) plagiarism, which is copying ideas, data, or words without attribution. Fanelli found that an average of 1.97% of scientists had admitted to having fabricated, falsified, or modified data or results at least once, whereas 33.7% had admitted to other questionable research practices. When asked about the behavior of their colleagues, 14.12% of scientists reported falsification of data and 72% reported other questionable research practices. As shockingly high as these rates of misconduct are, Fanelli claimed that "Considering that these surveys ask sensitive questions and have other limitations, it appears likely that this is a conservative estimate of the true prevalence of scientific misconduct."[22] Not many of these scientists would compare to Wakefield, but still these findings are worth keeping in mind. When scientists consciously commit fraud for whatever reason, things are bad, and it is difficult to deal with such cases. This is why journals nowadays ask authors to state potential conflicts of interest, as this is one way to see whether they would have reasons to commit fraud or be somehow biased.

However, scientists can also do a bad job without actually realizing it. Let us then consider some reasons for which this may happen, which all have to do with how scientists interpret the data available to them.

The suspended step from data to evidence

What scientists obtain from their studies are data. For instance, that the frequency of autism was x% among those children who were vaccinated and y% among those who were not are the data that scientists collect during their studies. As I explained in Chapter 1, data are just facts; they have no meaning on their own. When these data are analyzed, for instance used to make comparisons between two groups, they can become evidence, that is, a reason to arrive at a particular conclusion. Because scientists studying the same data can interpret them in different ways and draw different conclusions, data alone cannot directly point to particular conclusions. Data do not speak for themselves; rather, how they are interpreted is crucial. Furthermore, evidence can support a conclusion but can never prove it. But this does not diminish the validity of the conclusions. In contrast, it is exactly when scientists interpret the available data in the same way, and when these data become evidence for the same conclusions, that we can have trust in what they tell us.

However, scientists may have different personal perspectives and interpret data or explain the same phenomenon in different ways. Even the best scientists have their own ways of perceiving and interpreting the world. Conclusions are constructed based on the interpretations of facts that can vary depending on personal, idiosyncratic features. These may relate to previous experiences, worldviews, imagination and creativity, the sociocultural context in which one lives and works, and more. They may also depend on cognitive biases of different kinds as well as ambitions and motivations. Scientists are humans, and they can be subjective in their interpretations and conclusions. This subjectivity is often very good for science because different perspectives can result in discussions and debates that can lead to better understanding. Whereas it may be troublesome until scientists arrive at a consensus, it is also incredibly valuable as one might see what most others fail to see. When this consensus is reached, it means that most individual biases have been filtered (why this consensus matters was discussed in Chapter 6). But this subjectivity can also lead to problems.

Raphael Silberzahn and Eric Uhlmann conducted an interesting experiment. They recruited 29 teams of researchers and asked them to answer the following question: Are soccer referees more likely to give red cards to players with dark skin than to players with light skin? The researchers who agreed to participate in this experiment held different opinions about whether such an effect existed. They were all given the same large dataset from four major football leagues, which included referee calls, counts of how often referees encountered each player, and player demographics including the players' position on their teams, height, weight, and skin color. Each of the 29 teams of researchers developed its own method of analysis. All methods were shared with all researchers, anonymized and without revealing results, asking them to rate the validity of each method and to provide in-depth feedback on three of them. In the end, the teams were given the opportunity to revise their method based on the feedback received. Twenty of the twenty-nine teams found a statistically significant correlation between skin color and red cards. Dark-skinned players were found to be 1.3 times more likely than light-skinned players to receive red cards. However, the findings varied enormously, from a slight (and nonsignificant) tendency for referees to give more red cards to light-skinned players to a strong tendency of giving more red cards to dark-skinned players. Thus, an important conclusion was reached: "Any single team's results are strongly influenced by subjective choices during the analysis phase. Had any one of these 29 analyses come out as a single peer-reviewed publication, the conclusion could have ranged from no race bias in referee decisions to a huge bias." Figure 7.2, used in that article, provides a clear take-home message. Subjectivity exists, and we should keep it in mind. As the authors noted, "taking any single analysis too seriously could be a mistake, yet this is encouraged by our current system of scientific publishing and media coverage."[23] Rather than following individual studies, we should look where the field as a whole is going. This is why it is the consensus of the scientific community that matters.

Another important issue is what economist Gary Smith has described as "data torturing": the conscious or unconscious search for meaningful, or simply publishable, results. Two of the key problems Smith has identified are "HARKing," which means Hypothesizing After the Results are Known, and "p-hacking," which is "trying different combinations of variables, looking at

Figure 7.2 How the same data can be interpreted differently by different scientists. Reprinted with permission from Silberzahn, R., and Uhlmann, E. L. (2015). Crowdsourced research: Many hands make tight work. *Nature, 526*(7572), 189.

subsets of the data, discarding contradictory data, doing whatever it takes until something with a low p-value is discovered and then pretending that this is what you were looking for in the first place."[24] Let us consider these problems one by one.

The cartoon in Figure 7.3 brilliantly captures the concept of HARKing. A sharpshooter aims at the wooden fence and shoots several times with his firearm. Then he uses paint to draw targets wherever his bullets made a hole. In the end, it will look as if the sharpshooter had hit the center of all targets that were already there. But the truth is different: The sharpshooter did not really hit the center of the targets, because these were created once the holes were made. In a similar sense, scientists may develop a hypothesis after the evidence has been found (post hoc hypothesis) and present it as if it had been formed before the evidence was found (a priori hypothesis). Making inferences from data after the data were collected and presenting the inferences as hypotheses that had been formed right from the start is not a legitimate approach to do science. Social psychologist Norbert Kerr, who also coined the term, has proposed several reasons for which this is problematic, some of which are the following:

- Integrate false results into a theory.
- Overestimate the significance of the results.
- Do not communicate what did not work.
- Preclude the possibility of formulating a plausible alternative hypothesis.
- Adopt a theory that depends on the specific results and that is therefore not generalizable.
- Make others less receptive to serendipitous findings.
- Raise concerns about the validity and reliability of the scientific findings, as well as for the integrity of the scientists themselves.
- It is unethical not to communicate how one obtained the results they got.[25]

The scientifically legitimate approach is for researchers to develop hypotheses before collecting and analyzing data (see Box 1.1), so that the hypotheses are not affected by the results.

The other problem of "data torturing" is p-hacking: collect or select data or statistical analyses until significant results are obtained (Figure 7.4). A statistically significant result is one with a p-value lower than 0.05—which simply means that the probability for the relation identified to be due to chance is less than 5% or that it is more than 95% likely for the relation identified to be true. Some common practices of p-hacking are: (a) test a vast number of

Figure 7.3 HARKing or Hypothesizing After the Results are Known. In the same way that the sharpshooter draws a target around the holes after he made them so that it seems that they were there before he took the shots, scientists can develop hypotheses after the evidence is made available and present them as if they were made before the evidence was collected. Credit: Artwork by Dirk-Jan Hoek, concept by Eric-Jan Wagenmakers (CC-BY license).

outcomes and cherry-pick the ones that are statistically significant; (b) create new samples by adding more observations or eliminating those that do not support the hypothesis; and (c) try different statistical methods or models until a significant p-value is found. Unlikely as an outcome can be, it can materialize if a process is repeated many times, or if the sample is very large. If you perform an experiment multiple times, it is possible to arrive at a statistically significant result just by chance. Interestingly, p-hacking is possible to identify in the scientific literature. If researchers p-hack and turn a truly nonsignificant result into a significant one, then the p-curve's shape will be altered close to the perceived significance threshold, which is usually $p = 0.05$. As a result, a p-hacked p-curve will have an overabundance of p-values just below 0.05. In an analysis of a very large dataset that consists of p-values from different disciplines, it was shown empirically that p-hacking is widespread in the scientific literature. However, researchers also found that the effect of p-hacking is weak relative to the real effect sizes being measured.[26]

The effect size is actually a value that is a lot more important than p. The problem with p-hacking is finding a significant association where it

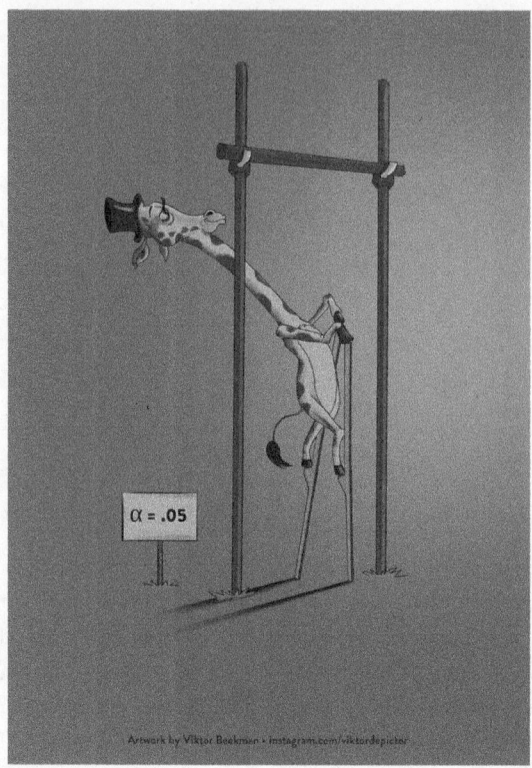

Figure 7.4 p-Hacking: When reaching the wished level is all that matters. The caricature represents how scientists may do whatever it takes to reach the 0.5 level of significance. Credit: Artwork by Viktor Beekman, concept by Eric-Jan Wagenmakers (CC-BY license).

does not really exist. But even if a significant association does exist, it does not tell much on its own. Rather, what matters is the size of the effect that this significant association has. Whereas a p-value can indicate whether an effect exists, it does not indicate how important the effect is. An example comes from the Physicians' Health Study impact of aspirin in preventing myocardial infarction. The research followed 22,000 people over an average of five years and found that the association between aspirin consumption and a reduction in myocardial infarction was highly statistically significant ($p < 0.00001$). However, the effect size was very small: a risk difference of 0.77%. As a result of that study, many people who were advised to take aspirin would not experience significant benefit and might actually risk adverse effects.[27] This is an example of how statistically significant associations might mislead.

The effect size is a statistical measure used to quantify the actual effects of a phenomenon, such as the impact of a treatment or intervention. The goal of research is to better understand the real-world significance of a phenomenon; estimating the statistical significance is not sufficient for this goal, as it is only about the improbability of the findings and not about their consequences. What matters more is the practical significance of the findings, and the effect size is a measure of this. Practical significance is inferred from the size of the effect, whereas statistical significance is inferred from the precision of the estimate. It is important to note that it is possible to observe an effect even though it is not statistically significant, something that can happen when the sample is small. When we are considering dichotomous outcomes (e.g., disease/not disease), then we can calculate the odds ratio: the odds that an outcome will occur given a particular exposure, compared to the odds of the outcome occurring in the absence of that exposure. When we are considering continuous outcomes, then we can calculate the difference in the average measures of two groups.[28] Let us consider the results of the 1954 Salk vaccine trials as an example. As described in Chapter 5, the goal was to evaluate the effects of the Salk poliomyelitis vaccine, and part of the trials consisted of the comparison of vaccinated children to those that had received a placebo (a salt solution). The calculation done is based on the data in Table 5.1, and its details are not necessary for our discussion here. What matters is that the effect size indicator that can be calculated is $r = 0.011$. The $r = 0.011$ may initially seem to be very small; however, it actually reflects a 1.1% decrease in paralytic polio, which is not insignificant.[29]

What is crucial to understand? (and also teach in schools!)

Two questions may have occurred to you by now. What counts as good science? And who decides what good science is? There are several different ways of evaluating the successes and failures of scientific research. But there exist some researchers who study science itself. One of them is John P. A. Ioannidis at Stanford University, who in 2005 published a now highly cited article with the title "Why most published research findings are false"(!) Strange as it might sound at first, this is a criticism that did not aim to take down science as an institution, but rather serve as a call to arms in order to improve it. Ioannidis pointed out that in many cases, there has been no attempt to confirm the findings of research studies by means of repeating them (Box 7.1). As a result, conclusions are made on the basis of individual studies that may contradict one another. What is needed from researchers, therefore, is to look at what various studies on the same question are finding.

Box 7.1 Repeating scientific studies

One of the ways to confirm the validity of new scientific findings is by repeating the research that produced them. When a scientific study fails to independently confirm the findings of a previous one, there are reasons to be concerned. Is this due to a lack of rigor, or is it simply due to the nature of the new findings? Whatever the case, independent confirmation of findings is essential for having confidence and trust in them.

There are two distinct features of such good science:

Reproducibility: Obtain consistent results with the initial study using the same data, computational steps, methods, and conditions of analysis.

Replicability: Obtain consistent results across different studies aimed at answering the same scientific question, each of which has obtained new data.

Being able to reproduce the results of previous studies with the same data or replicate the results of a previous study with new data is considered a hallmark of good science, which supports its self-correcting nature. However, a successful replication does not guarantee that the original scientific results of a study were correct, nor does a single failed replication conclusively refute them.[31]

Ioannidis has suggested that it is impossible to be 100% certain about what the truth is in any research question. However, particular features of a scientific field can make research findings more likely to be true when researchers:

1. conduct studies with thousands of participants rather than with one or a few hundred participants;
2. have large effects and higher relative risks, such as the impact of smoking on cancer or cardiovascular disease, rather than having smaller effects and relative risks, such as genetic risk factors for complex diseases;
3. rely on meta-analyses (studies that analyze data from independent studies of the same topic in order to determine overall trends) rather than on initial studies;

4. conduct studies with lower flexibility in designs, definitions, outcomes, and analytical modes, which limit the potential for a biased interpretation of results by interpreting what would be "negative" results as "positive";
5. lack financial interests;
6. are not committed to a particular view or wanting to defend one's personal findings;
7. have many research groups involved in research on the same topic, so that they try to confirm or refute the findings of other groups, and thus examine why different results were reached when different methods were used.[30]

Besides asking whether research findings are true, another crucial question is whether they are useful. For instance, what is the clinical significance of research findings in medical practice? Clinical research is useful when it can lead to a favorable change in decision-making, either by itself or when integrated with other studies and evidence. There are particular questions, the positive responses to which are indicators of useful clinical research, such as:

- Is there a health problem that is important enough to fix?
- Has prior evidence been systematically assessed to inform new studies?
- Is the proposed study large and long enough to be sufficiently informative?
- Does the research reflect real life?
- Does the research reflect the top priorities of patients?
- Is the research worth the money spent?
- Can this research be done?
- Are methods, data, and analyses verifiable and unbiased?

Ioannidis has argued that many studies, even those published in prestigious medical journals, do not positively answer these questions, and in fact, only a few satisfy all of them. This is why medical research may fail to be useful.[32]

Does all this look pessimistic? Well, perhaps it does, but it is actually about being realistic. As Ioannidis put it:

Our society will benefit from using the best available science for governmental regulation and policy. One can only applaud when governments want to support the best possible science, invest in it, find ways to reduce biases, and provide incentives that bolster transparency, reproducibility, and the application of best methods to address questions that matter. However, perceived perfection is not a characteristic of science, but of dogma. Even the strongest science may have imperfections. In using scientific information for decision-making, it is essential to examine evidence in its totality, recognize its relative strengths and weaknesses, and make the best judgment based on what is available.[33]

In short, scientific knowledge and understanding are not absolute or fixed. Rather, they are incomplete and, in some sense, tentative. Science has an inherent self-correcting nature, through its internal procedures of review and interrogation, which constantly question much of what scientists know. Not all parts of a theory are equally subject to change because some are so well supported that it would take a tremendous amount of conflicting evidence to warrant changing them, whereas other parts are not as well supported and so can be changed on the basis of less evidence. But changes do occur once scientists come across conflicting evidence. This tentativeness and openness to change may seem like a weak aspect of science, a kind of Achilles' heel. But this is not the case at all. In contrast, it is the self-correcting nature of science and scientists' awareness of issues such as those considered in this section that guarantees the production of solid knowledge and understanding.

Given the daily headlines about individual scientific studies that found one or the other thing, it is best to ignore them and try to consider the totality of the evidence, that is, the findings of all research groups and not only the statistically significant findings of one or a few of them. In the case of complex diseases such as cancers, conclusions are difficult to make because finding associations is not necessarily informative. For instance, associations with cancer risk or benefits have been found for most food ingredients (Figure 7.5). A study aimed at examining the conclusions, statistical significance, and reproducibility of published studies on associations between specific foods and cancer risk. Fifty common food ingredients were selected, and articles about their cancer risk were found for 40 of them. From 264 single studies, 103 studies concluded that the ingredient tested was associated with an increased risk and 88 studies concluded that the ingredient tested was associated with a decreased risk. However, 36 meta-analyses presented different results, with only 4 of them reporting an increased risk and 9 of them reporting a decreased risk. It was thus concluded that multiple studies are necessary for scientists to draw valid conclusions, and caution is required when drawing conclusions until strong evidence becomes available.[34]

How we should think about scientific studies can be aptly summarized by a metaphor. When someone is too involved in the details of a problem, one may lose sight of the problem as a whole. In such cases, we say that this person

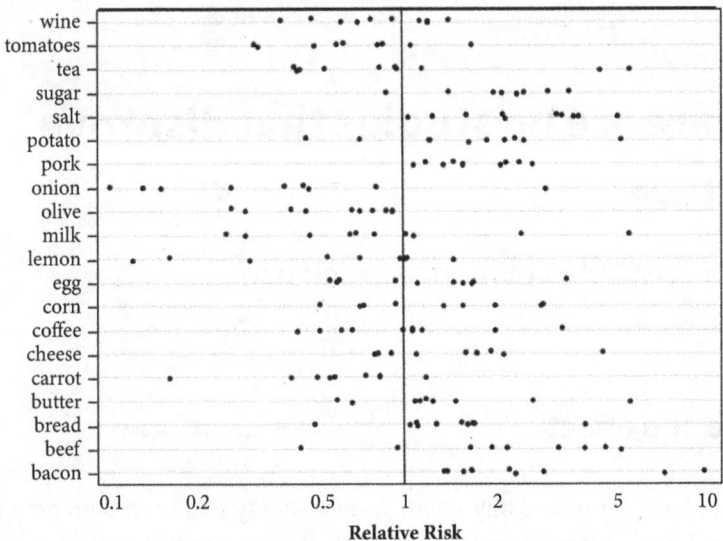

Figure 7.5 Effect estimates reported in the literature by food ingredient (only ingredients with 10 or more studies are shown). Individual studies point to contradictory conclusions about the relation between food ingredients and cancer. This is why one should look at meta-analyses and many studies together, rather than individual ones. Reprinted by permission from Schoenfeld, J. D., & Ioannidis, J. P. (2013). Is everything we eat associated with cancer? A systematic cookbook review. *American Journal of Clinical Nutrition, 97*(1), 127–134.

"misses the forest for the trees," because that person has focused too much on the details (the trees), missing the broader picture (the forest). We can thus think of the individual studies within a particular scientific discipline, each addressing a particular aspect of the broader research program, as trees that together form an interconnected forest. You can imagine Figure 7.5 as a map of this forest. What we should do is to look at the general picture, for example, that most studies show an increased risk for cancer because of eating bacon, and no risk from eating olives. Focusing on one of these studies would be like missing the forest for the trees.

8

"There are no studies that disprove it either"

What constitutes evidence in science

Cause and effect

> ... I have yet to find any scientist who will say that there's no doubt, no doubt, that the mercury in vaccines does not contribute to autism. Now, they'll say there's no scientific evidence, there's no studies or anything that proves that yet. But turn that around, there are no studies that disprove it either. And so they're skirting the issue.[1]

It was December 10, 2002, during a hearing in the House of Representatives before the Committee on Government Reform. These are the words of Dan Burton, a Republican politician who was a US Representative for Indiana between 1983 and 2013. The topic was "Vaccines and the Autism Epidemic: Reviewing the Federal Government's Track Record and Charting a Course for The Future." Burton argued:

> ... my grandson and thousands of children across this country were normal kids and they got vaccinated with multiple vaccines. And mercury in the brain has a cumulative effect; all scientists will tell you that it doesn't wash out easily. It gets in the fatty tissues and it stays there so it has a cumulative effect. And yet we continue to get reports that say there's no scientific evidence that mercury causes autism. They don't say it doesn't, they say we can't conclusively prove that mercury causes autism. They don't say it doesn't.[2]

Let us unpack Burton's argument.

Burton and others, including many parents whose children have autism, blame vaccines for that. As already mentioned, it can be the case that the first symptoms of autism appear after vaccination. Therefore, Burton and others argue, vaccines are the cause of autism. The idea here is that when something happens right after something else, the latter is the cause of the former.

Trusting Science. Kostas Kampourakis, Oxford University Press. © Oxford University Press (2025). DOI: 10.1093/oso/9780197787106.003.0008

They also specify the ingredient responsible for this: mercury. According to the Centers for Disease Control (CDC) in the USA, this is a natural element, and people are likely to be exposed to two types of it: methylmercury and ethylmercury. The first is found in particular kinds of fish, and high exposure levels can be toxic. The second is contained in thimerosal, which was used for decades in the USA as a preservative in multidose vials of vaccines. The reason for this is that needles that are inserted in vials that are used more than once can introduce fungi and bacteria therein, and so those can in turn infect the person who will subsequently receive a vaccine dose from the same vial. Thimerosal serves to prevent the growth of fungi and bacteria inside the vial. When it is inserted in the human body during vaccination, it is quickly broken down to ethylmercury and thiosalicylate, which are in turn quickly eliminated. As a result, mercury is not expected to accumulate in the body and cause harm, contrary to what Burton claimed above.

The official information that a citizen gets from CDC is this: "Scientific research does not show a connection between thimerosal and autism."[3] This is exactly the statement that annoys Burton. What this statement means is that in trying to figure out if there is a cause–effect relation between thimerosal (cause) and autism (effect), scientists have not (so far) found any evidence that such a relation exists. In short, scientists have concluded that there is *absence of evidence* that thimerosal causes autism. There are at least two ways to arrive at such a conclusion.

One is to compare a group of children who have received thimerosal to another group who have not received it. If the incidence of autism, that is, the number of children who develop autism during a particular period, is significantly higher in the group who received thimerosal than in the one who did not, then one can infer an association between thimerosal and autism. This association in turn might, but also might not, be due to a causal effect of the thimerosal in the emergence of autism. In other words, the association between thimerosal and the higher incidence of autism would provide evidence that thimerosal *might* contribute to autism (I return to this *might* below). If, however, the incidence of autism was more or less the same in the two groups, then there would be no reason to think that there is an association between thimerosal and autism. This is exactly what a study with 467,450 children found. The researchers compared children who received a vaccine containing thimerosal with children who received a thimerosal-free formulation of the same vaccine. The risk of autism was not significantly different between the two groups. Therefore, these results did not support a cause–effect relation between thimerosal and autism.[4]

Another way to figure out if there is a cause–effect relation between thimerosal and autism is to compare children with autism to matched controls (that is, children of the same age and gender without autism), with respect to their exposure to ethylmercury at particular ages. If it were found that children with autism had been exposed to ethylmercury levels that were significantly higher than those of the controls, this would provide evidence that ethylmercury *might* contribute to autism (again, I return to this *might* below). If on the other hand the children in both groups had been exposed to similar levels of ethylmercury at particular ages, then there would be no reason to believe that ethylmercury contributes to autism. In such a case, one could say that there was no evidence for a causal connection between them. This is indeed what such a study of 256 children with autism and 752 controls has found. On average, the children in the two groups had similar cumulative ethylmercury exposures at the end of each period studied (prenatal; birth to 1 month old; birth to 7 months old; birth to 20 months old).[5]

Let us consider now the *might* I used above. I wrote that even if it were found that children with autism had been exposed to ethylmercury levels that were significantly higher than those of the controls, this would provide evidence that ethylmercury *might* contribute to autism; this would not be evidence that ethylmercury *did* contribute to autism. The reason is that a case-control study like the ones I have just described can only provide a statistical association between two variables. A cause and an effect are always statistically associated, because one causes the other; therefore, the more we have of the cause, the more we also have of the effect. However, the opposite is not necessarily true: two variables that are statistically associated do not necessarily have a cause–effect relation. Statistical associations can also exist between two variables that are not directly related to each other, because they are both related to a third variable. The classic example is the statistical association between eating ice cream and drowning incidents. Such an association does not entail that eating ice cream causes drowning. Rather, the association between eating ice cream and drowning incidents exists because they have a common cause: a warm summer weather. The warmer the weather is, the more likely it is that people will eat ice cream or go swimming, and the more people swim, the more likely is that someone will drown.[6]

Returning to the lack of association between exposure to mercury and autism described above, you might think that these are single studies and might thus not be adequate, and you would be absolutely right! As explained in the previous chapter, no single study in science is ever adequate to establish anything. This is why it is important not to look at individual studies separately, but at all (or most) of the studies in a field together. In order to better understand the possible association between childhood vaccinations and the subsequent development of autism, a group of researchers performed a meta-analysis. In particular, they looked at two kinds of studies: case-control studies, such as the ones described above, and cohort studies (studies that follow a particular group of people—a cohort—who share a particular characteristic for a particular period). They thus analyzed data from five cohort studies involving 1,256,407 children, and five case-control studies involving 9,920 children. Overall, no statistical association was found between vaccination and autism.[7] With meta-analyses like this one, we can more confidently state that there is no evidence for a contribution of vaccines (or thimerosal, or mercury) to the development of autism, and that there is *absence of evidence* for such a relation.

I should note at this point that the best way to establish a causal relationship, or the lack thereof, is with controlled experiments (see also Chapter 5). If experimentation with humans were possible—it is good that it is not!—we could design a randomized controlled trial to test this. We could have numerous children, matched for age, gender, and any other variable we wanted, randomly assigned to one of two groups: one that would receive, say, thimerosal (the experimental group), and another that would not receive it (the control group). By administering thimerosal to children at exactly the same age, and by keeping all the other variables the same, we could follow these children and see if they developed autism or not. If significantly more children who had received thimerosal than those who had not developed autism, we would have strong evidence that thimerosal *might* (again, for the reasons described above) cause autism. But if no differences were found between the two groups, this would be *evidence of absence* of a contribution of thimerosal to autism. Then to better understand the pathology of autism, we might even try to study in those children who developed autism the cellular or protein changes in the cerebellum, hippocampus, or amygdala—the three regions of the brain that are commonly found to have such changes in the brains of people with autism after their death. As this would require dissections, I assume that neither Burton nor anyone else would like to see such a procedure take place. However, a similar study with infant rhesus macaques who received vaccines containing thimerosal found neither behavioral nor neuropathological differences between the experimental and the control groups.[8]

We have seen how scientific studies do not support that thimerosal causes autism. However, we have not yet addressed Burton's concern. Yes, there are studies that do not "prove" such an association—he admitted that. But how about studies that directly "disprove" it? It is useful to note here that in science we cannot really "prove" anything. We can only find evidence to support a hypothesis, or to arrive at a conclusion. But there is no proof of anything in any absolute sense. Having clarified this, Burton's concern is still there. He is not satisfied with the *absence of evidence* for a connection between vaccines and autism; what he is looking for is something different: *evidence of absence* for a connection between vaccines and autism. So, what Burton is looking for are not studies that cannot find an association, but rather studies that can show that such an association does not exist. Is this possible?

Absence of evidence vs. *evidence of absence*: the question about the existence of god

To address Burton's concern, we need to answer two distinct questions:

(Q1) Is the *absence of evidence* for a phenomenon also evidence of its absence?

(Q2) Can we find evidence for the absence of a phenomenon?

Let us begin with the first question, considering a very different issue: the question of the existence of god. Evolutionary biologist Jerry Coyne has argued that:

> With the notion of theistic god and a vernacular notion of "proof" in hand, we can disprove a god's existence in this way: *If a thing is claimed to exist, and its existence has consequences, then the absence of those consequences is evidence against the existence of the thing.* In other words, the absence of evidence—*if evidence should be there*—is indeed evidence of absence ... Many gods claimed to exist should have observable effects on the world. ... But the evidence isn't there: we see no miracles or miracle cures in today's world, much less any wondrous signs of a God who presumably wants us to know him.[9]

Coyne thus makes the argument that if we do not find evidence for the existence of god, then this entails that god does not exist: The *absence of evidence* for the existence of god is also *evidence of god's absence*. I respectfully disagree. The fact that we have not observed the actions of an all-powerful god does not entail that such a god does not exist. One cannot establish the nonexistence of something simply because there is no evidence of its existence. Such evidence may actually exist, but it may have simply escaped our attention. Coyne's argument is actually a slippery one that he himself might not agree with in other contexts. Anti-evolutionists might cite him, arguing that the *absence of evidence* for a particular transition in evolution is also evidence that this transition never happened. They could do this for every single piece of evidence that is not currently available. Just replace god in the quotation above with any evolutionary transition you want. However, anti-evolutionists would be wrong. That we have not found evidence for the existence of a transitional form in evolution does not entail that such a form has not existed. It is possible that it in fact existed, but we have not found the relevant evidence yet. Therefore, the *absence of evidence* is not the *evidence of absence*.

Consider the evolution of tetrapods (four-limbed animals) from fish (discussed in detail in Chapter 10). This is not a self-evident evolutionary transition. However, finding *Tiktaalik*, which has several features intermediate between fish and tetrapods, has shown that such a transition could have been possible (Figure 10.4). Here is what anti-evolutionists could have said before the discovery of *Tiktaalik*, paraphrasing and citing Coyne:

> we can disprove the evolution of tetrapods from fish in this way: *If a thing is claimed to exist, and its existence has consequences, then the absence of those consequences is evidence against the existence of the thing.* In other words, the absence of evidence—*if evidence should be there*—is indeed evidence of absence ... The evolution of tetrapods from fish should have observable effects on the world. ... But the evidence isn't there: we see no transitional forms of this kind.

But this would be absurd, because a transitional form that has not yet been found may eventually be found. And in the case of the evolution of tetrapods from fish, we now have a good sense of what these forms might look like (see Chapter 10).

Having now explained why the *absence of evidence* for a phenomenon is not generally evidence of its absence, let us now consider whether we can find evidence for the absence of a phenomenon. At first sight, Burton's request for studies that show that thimerosal does not cause autism might seem reasonable. A similar request was made in 2017 by Robert F. Kennedy Jr. and Robert De Niro. Robert F. Kennedy Jr. is the son of US Attorney General Robert F. Kennedy and the nephew of President John F. Kennedy, both assassinated during the 1960s. He is also the editor of a book titled "Thimerosal: Let the Science Speak. The Evidence Supporting the Immediate Removal of Mercury—a Known Neurotoxin—from Vaccines." In 2005, he published an article (now removed) under the title *Deadly Immunity* in the Rolling Stone and Salon magazines, arguing about a massive conspiracy involving thimerosal.[10] Robert De Niro is a famous actor who is not as outspoken a critic of vaccines as Kennedy is, but who is concerned because he has a son with autism, Elliot. Here is their statement:

> We hereby issue a challenge to American journalists (and others) who have been assuring the public about the safety of mercury in vaccines. We will pay $100,000 to the first journalist, or other individual, who can point to a peer-reviewed scientific study demonstrating that thimerosal is safe in the amounts contained in vaccines currently being administered to American children and pregnant women.[11]

One can easily dismiss this request as anti-vaxxer nonsense. However, what Burton, Kennedy, and De Niro are asking for is certainly something everyone would like to know. Therefore, it is worth considering whether what they are asking for is also feasible. The short answer is that it is not. What they are asking for is not possible for science to offer, because it is in principle impossible. Let us see why.

Imagine that the FBI enters your home and arrests you under the charge that you are Batman. Their problem is that during the night you fight crime, even though you have no jurisdiction to do so. This is why the police exist, and nobody else—including Batman—is allowed to do this (forget for now the films and the good relation between Batman and the police commissioner). So, you are taken in for interrogation. Several hours later, your lawyer tells you that the FBI did not find anything in your home that might incriminate you: no Batman uniform, no Batmobile, nothing. There is no evidence whatsoever that you are Batman—this is *absence of evidence*. Whereas you are frustrated that you had to go through all that, you feel relieved that you can at least go home to your family. But no! The FBI tells you that you cannot go because, as I have also argued here, *absence of evidence* is not *evidence of absence*. You may be Batman, and it just happens that they have not been able to find your equipment so far. Therefore, you have to stay in custody until the FBI finds evidence that clearly disproves that you are Batman.

If you find this argument absurd, then what Burton and the others are asking for should be considered absurd too. Here are the key points:

- It is possible to find something that shows that you are Batman or that thimerosal causes autism. This is *evidence*, and it can be found in various ways such as those described above.
- It is also possible to find no evidence that you are Batman or that thimerosal causes autism. This is *absence of evidence*. This does not definitively mean that you are not Batman or that thimerosal does not cause autism. Perhaps both are true, but we have not yet found the

relevant evidence. But insofar as this is the case, all we can say is that
the available evidence does not support these conclusions.

- It is nearly impossible to find evidence that you are not Batman or that
  thimerosal does not cause autism. This is *evidence of absence*. The reason
  for this is that it is not possible to find evidence for something that does
  not occur. This is why Burton's argument fails.

There is only one way that we can have *evidence of absence* based on the
absence of evidence. Imagine that you hear what sounds like a dog barking in
the backyard. Once this happens, you have reasons to believe, based on your
background knowledge and previous experiences, that there will be evidence
of a dog being in the backyard, such as visual evidence of the dog, paw prints,
and so on. In such a case, if you were to go to the yard and discover a complete
absence of evidence of a dog, that is, you don't see any dog, any paw prints,
and so on, that would be evidence that there wasn't a dog in the yard, and
that you heard something else. For all this to happen, it is necessary to have
reasons to expect to find evidence of the presence of a dog and not an absence
of such evidence. This can be summarized as follows:

1. If a dog had been in the backyard, there wouldn't be an *absence of
   evidence* of its presence (not seeing any dog or paw prints, etc.).
2. There is an *absence of evidence*.
3. Therefore, there was no dog in the backyard.

This can be generalized as follows:

1. If X were the case, there wouldn't be an *absence of evidence*.
2. There is an *absence of evidence*.
3. Therefore, X isn't the case.[12]

And it can also be applied to the case of thimerosal:

1. If thimerosal caused autism, there wouldn't be an *absence of evidence* of
   its presence (no statistical association between the two).
2. There is an *absence of evidence*.
3. Therefore, thimerosal does not cause autism.

I should note at this point that these arguments are valid because we know
how dogs look like and behave, as well as why a substance containing mercury
might cause harm to human health (methylmercury is toxic, so one might
suspect that ethylmercury could be toxic too, even though it does not seem

to be). However, we cannot apply the above scheme to Coyne's argument about god, or to the FBI's argument about you being Batman, because we do not have any previous experience of what to expect as evidence for god or Batman.

Let us now consider in more detail what scientific evidence is.

What scientific evidence is

In a very accurate account of the issues related to vaccination in an episode of *Last Week Tonight*, John Oliver criticized Burton's argument by stating "... proving a negative is an impossible standard ..." and that "... when a scientist says we have no evidence that there's a link between vaccines and autism what they're really saying is you know we are as positive as someone can humanly be that there's no link ..."[13] Science is a quest for understanding, and also the most rational way to approach the natural world. At the same time, however, science cannot provide definitive answers. Each time we can only arrive at the best possible explanation, given the available evidence. Therefore, whenever people deal with scientific issues, they need to keep in mind some specific features of science as a way of acquiring understanding, such as those I presented at the beginning of Chapter 1. What these entail is that scientists can rely on empirical evidence for arriving at conclusions that are most likely to be accurate.

Let us consider how the CDC summarizes the current knowledge regarding thimerosal and autism:

Scientific research does not show a connection between thimerosal and autism.

Research does not show any link between thimerosal in vaccines and autism, a neurodevelopmental disorder. Many well conducted studies have concluded that thimerosal in vaccines does not contribute to the development of autism. Even after thimerosal was removed from almost all childhood vaccines, autism rates continued to increase, which is the opposite of what would be expected if thimerosal caused autism.[14]

What is mentioned in the heading and in the first two sentences has already been discussed. The last sentence, however, is particularly interesting. Thimerosal was taken out of childhood vaccines in the USA in 2001. The only vaccines that currently contain thimerosal are the multidose influenza (flu) vaccines, but there also exist alternatives without thimerosal. This simply entails that Burton, Kennedy, and De Niro are fighting a strawman; they are concerned about the effect of thimerosal contained in vaccines, even though almost all vaccines do not contain thimerosal!

Equally interesting is the point made that if thimerosal was indeed the cause of autism, once it was taken out from vaccines, one should have expected to see fewer cases of autism than before—because the causal factor would no longer be there. However, not only was this not the case, but also the cases of autism have increased since then! Whereas in 2000 the prevalence of autism (the proportion of children who were diagnosed with autism in a given time period) was 6.7 per 1,000 children, in 2018, it was 23 per 1,000 children—more than a threefold increase! The reason for this is not that there is now more autism than before. Rather, what has changed is that more children with autism get to be diagnosed. For instance, in 2004, there were significant differences among different US-census social groups with respect to the diagnosis of autism. In 2018, there were no such differences, which means that during the intervening years, those groups that previously had less access to healthcare services may now have more access to them.[15]

This brings up another important issue: how data become evidence for a conclusion depends largely on the context, and one has to be aware of that. It would be easy for vaccine skeptics to blame vaccines for the increase in the prevalence of autism. Indeed, former US president Donald Trump tweeted in 2012 (when he was not yet President) that "A study says @Autism is out of control—a 78% increase in 10 years. Stop giving monstrous combined vaccinations immediately. Space out small individual shots—small babies can't handle massive doses. Get smart -and fast- before it is too late."[16] Trump was apparently referring to a report that stated that:

> Comparison of 2008 findings with those for earlier surveillance years indicated an increase in estimated ASD prevalence of 23% when the 2008 data were compared with the data for 2006 (from 9.0 per 1,000 children aged 8 years in 2006 to 11.0 in 2008 for the 11 sites that provided data for both surveillance years) and an estimated increase of 78% when the 2008 data were compared with the data for 2002 (from 6.4 per 1,000 children aged 8 years in 2002 to 11.4 in 2008 for the 13 sites that provided data for both surveillance years).[17]

A 23% increase from 2006 to 2008, and a 78% increase from 2002 to 2008! This certainly looks like a lot, but it happened for the reasons I have just explained—it is not due to vaccines as Trump mistakenly claimed. An increase is an increase, but one has to be aware of the context in order to understand why it happened, and what it entails. There was an increase because more children than before were diagnosed with autism, not because more children than before developed autism.

As already explained, for data to become evidence, we need to be aware of much more than the facts themselves. The 78% increase is a fact—data. But these data do not automatically become evidence for the contribution of vaccines to the development of autism as Trump claimed. Neither can we find evidence that thimerosal does not cause autism as Burton required. We can only find data about the relation between these two, and so far we can only say that there is no evidence for it. The real problem here is not thimerosal or vaccines, but that Burton and others lack a basic understanding of what evidence is in science.

What is crucial to understand? (and also teach in schools!)

The key issue considered in this chapter is that, aside from some very special cases, we cannot have evidence for the absence of a phenomenon, but only *absence of evidence*. But even if we found evidence for a correlation between two variables, it would not necessarily entail causation. Let us consider some examples. In 1951, physician Benjamin Sandler, at the time a nutrition specialist at the Oteen Veterans Administration Hospital, published a short book in which he argued, providing what he considered as the relevant evidence, that a major cause for susceptibility to polio was low blood sugar in the human body.[18] Sandler suggested that a normal blood sugar content of 100 milligrams in each 100 cubic centimeters of blood was necessary to maintain resistance to infection by polio. If this blood sugar concentration was reduced significantly (for instance, from 75 to 55 mg.), this could lower the body's defenses and allow infection by the poliovirus. Sandler argued that this lowering of blood sugar was caused, perhaps paradoxically, by eating too much sugary and starchy food. Therefore, he believed, the best preventive measure for polio would be a diet low in sugars and starches. Once this was taken up, according to Sandler, the body could build up sufficient resistance to the poliovirus within 24 hours, thus preventing disease. To have a more long-term protection, one should continue the diet.

Ten years before, Sandler had conducted experiments with rabbits, and his conclusion had been that the virus could grow and infect cells only when the blood sugar level in the host organism was kept low by administering insulin. When blood sugar was left to return to its normal levels, the virus was somehow prevented from growing and infecting as well as previously. Therefore, Sandler inferred that low sugar levels might have an important role in determining susceptibility to the poliovirus. He also speculated that the lack of metabolic reactions in cells due to low sugar levels somehow rendered them susceptible to the poliovirus.[19] Subsequently, Sandler encountered polio patients, many of whom had a diet deficient in protein and rich in the cheaper starchy foods.

Sandler thus felt that he had the solution to the polio epidemic. As he wrote in 1951, "I hoped that publication of my experiments in the medical journal in 1941 would stir public health authorities to explore the possibilities implicit in my experimental results. However, nothing ever came of the publication. The data lay buried on the bookshelf. And so I decided to assume personal responsibility by informing the public. It was a bold step and required some courage because a professional career could have been jeopardized. Looking back, I am happy I took the step." In the summer of 1948, he seized the opportunity to do something about it. At the time, he was living in Asheville, NC, a city of 55,000 people. In June, it had become clear that a major polio epidemic was upcoming. In July, the city took measures: closures of public spaces such as churches, theaters, swimming pools, parks, and recreation areas, whereas public gatherings were discouraged, and children were kept home.

On August 1, Sandler approached the local newspapers and informed them about his experiments. Thus, on August 4, 1948, *The Asheville Times* published a detailed account of his experiments, along with recommendations to refrain from consuming foods with sugars of all kinds, as well as from eating meat or dairy products, and to rest as much as possible. According to Sandler, the results were stunning. Whereas the number of cases in 1948 was higher than the respective number in 1946 until July 31, for the next six weeks, the number of cases was much lower, and this was a decline that had not been observed in previous polio epidemics. As Sandler wrote: "The break in the 1948 curve occurred during the week ending August 7 and coincided with the release of the diet story on August 4 and 5. This immediate effect need not be surprising since it was stated 'without reserve' that strict adherence to the diet would afford protection within 24 hours, because the change in diet has an immediate effect on blood sugar levels." Sandler's conclusion, based on a "conservative estimate," was that the diet campaign resulted in around 3,000 fewer polio cases than what was expected between August 7 and September 11, 1948, and 5,000 fewer cases until the end of the year.[20]

Not everyone was as thrilled as Sandler was with these results. On August 9, 1948, the newspaper *The Durham Sun* reported that James H. Cherry, president of the Buncombe County Medical Society argued that the consensus among the local physicians was that there was no justification for Sandler's recommendations.[21] The fact that the cases of polio declined after the diet campaign did not necessarily entail that children's low blood sugar protected them from polio for various reasons. First, one could not even know if families implemented Sandler's dietary recommendations, insofar as there were no detailed records of this. The least that Sandler should have done would have been to monitor the diet of several families for some time, both those who implemented his recommendations and those who did not, and then compare the number of polio cases in both groups. But even if he had done this, there would still be methodological questions. That the cases of polio declined after the diet campaign does not in any way entail that the latter was the cause of the former. It could have simply been a coincidence producing a temporal correlation. To establish causation, a lot more data than what Sandler had available would be required.

Let us consider another example. An article in the prestigious journal *The New England Journal of Medicine* explored whether there was a correlation between a country's chocolate consumption and its population's cognitive function, the latter measured by the number of Nobel Prizes per capita. The assumption was that chocolate consumption could hypothetically improve cognitive function in whole populations. The author of this article, Franz H. Messerli, MD, from St. Luke's-Roosevelt Hospital and Columbia University, New York—who by the way reported "regular daily chocolate consumption, mostly but not exclusively in the form of Lindt's dark varieties"—found the following results:

> There was a close, significant linear correlation ($r = 0.791$, $P<0.0001$) between chocolate consumption per capita and the number of Nobel laureates per 10 million persons in a total of 23 countries (Fig. 1). When recalculated with the exclusion of Sweden, the correlation coefficient increased to 0.862. Switzerland was the top performer in terms of both the number of Nobel laureates and chocolate consumption.[22]

His conclusions speak for themselves:

> Chocolate consumption enhances cognitive function, which is a sine qua non for winning the Nobel Prize, and it closely correlates with the number of Nobel laureates in each country. It remains to be determined whether the consumption of chocolate is the underlying mechanism for the observed association with improved cognitive function.[23]

· Needless to say, other scientists expressed their concerns about these results on methodological, statistical, and logical grounds. Even though a beneficial effect of chocolate consumption on cognitive function is plausible, and more logical than the opposite effect of Nobel Prize awards increasing the consumption of chocolate, at the statistical level, correlation never implies causation on its own. As Messerli himself pointed out in his conclusion, finding the underlying biological mechanism would be crucial. Most importantly, the key problem with his interpretation was that he arrived at a conclusion about individual behaviors based on data from group behaviors without any reason to think that the relationships observed at the group level necessarily hold for individuals. It is one thing to measure the country-averaged chocolate consumption and another to measure the actual consumption of Nobel laureates themselves, and it is perhaps the latter that might provide evidence for the relation between chocolate consumption and cognitive performance (something also acknowledged by Messerli himself in his discussion of the limitations of his analysis).[24] Whether Messerli believed in what he wrote or he was just joking is open to investigation.

Another more recent example is about the cause of COVID-19. During the pandemic, concerns were expressed that the disease was caused by the 5G antennas. The fact that the denser the 5G network was, the more COVID-19 cases there were, was for many people evidence that the former was the cause of the latter. Maps were used to show the correspondence between the distribution of COVID-19 cases and the distribution of 5G antennas.[25] Astonishingly, 5G antenna engineers were threatened or attacked because of this perceived connection.[26] In my view, they were just unlucky; Taco Bell restaurant employees could have been attacked too, because as it happens the distribution of COVID-19 cases and Taco Bell restaurants in USA was also quite similar. But Taco Bell employees were lucky. And I mean this, because as you can imagine, the similarity in the distribution of COVID-19 cases and 5G antennas or Taco Bell restaurants cannot be coincidental. And indeed, it is not, but it is not due to a causal relation between them. Neither 5G antennas nor Taco Bell cause COVID-19. There is another reason for their common distribution, which is population density. COVID-19 cases tend to be higher in densely populated areas where people live and work in close proximity, making it easier for the virus to spread. These densely populated areas are also where there is a higher demand for mobile connectivity and fast-food services, leading to a higher concentration of 5G antennas and Taco Bell restaurants. Therefore, the overlap in the distribution of coronavirus cases with 5G antennas and Taco Bell restaurants can be explained by the common factor of high population density rather than any direct connection between them.

To sum up: We have evidence for a causal connection between A and B only when we know the mechanism whereby A causes B. A first step towards getting there is to find statistical associations between the presence of A and the presence of B. This might point to a causal connection between A and B, but it does not have to be so. And if we do not find any statistical association between A and B, as in the case of thimerosal and autism, then we have no reason to think that A causes B. In this case, we do not need to have *evidence of absence*; the *absence of evidence* is enough.

That the importance of correlations can be exaggerated even though they do not entail causation has been brilliantly shown by Tyler Vigen in his book and website "Spurious Correlations." Vigen has shown that we can find statistical associations between, say, US spending on science space and technology on the one hand and suicides by hanging, strangulation, and suffocation on the other (Figure 8.1a). Or we can find a statistical association between per capita consumption of mozzarella cheese and civil engineering doctorates

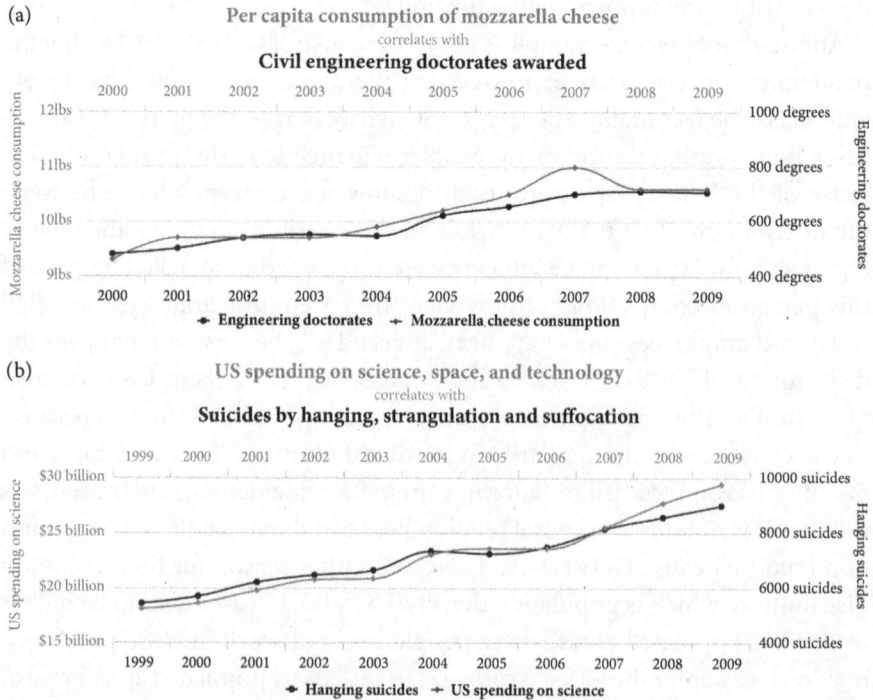

Figure 8.1 Spurious correlations. Graphs like these seem to show an association between entirely unrelated variables, such as US spending on science, space, technology and suicides by hanging, strangulation, suffocation, or between per capita consumption of mozzarella cheese and civil engineering doctorates awarded. However, none of these correlations represents any cause–effect relation.

awarded (Figure 8.1b). These correlations are obviously not due to real cause and effect relations, but mostly due to the ways the respective graphs were produced. Therefore, we should pay attention not only to whether the lines in the graphs align well, as is the case in the graphs of Figure 8.1, but also to the values on the various axes. More specifically:

- Y-axis scales that measure different values may show similar curves that shouldn't be paired.
- Even when the Y axes measure the same category, changing the scales can alter the lines to suggest a correlation.
- Plotting unrelated datasets together can make it seem that changes in one variable are causing changes in the other.[27]

9

"The government received our petition and immediately conceded. A vaccine injury had caused permanent brain damage."

Political decisions and their implications

Vaccine Injury

> Terry Poling: "Something happened after the vaccines. She just deteriorated and never came back."
>
> Jon Poling: "Vaccines are one of the most important advances in the history of medicine, but people need to know there is a risk to every medicine. There may be a small percentage of people who are susceptible to injury. I think we need a grassroots movement among pediatricians to be more conservative, and not give so many shots at once."[1]

Meet the Polings, parents of Hannah. Terry is Hannah's mother, a registered nurse and an attorney. Jon is Hannah's father, a neurologist with a PhD in biophysics. On March 6, 2008, they gave a news conference in Atlanta, after government health officials conceded that vaccines worsened a rare, underlying disorder that ultimately led to autism-like symptoms in Hannah, and that she should be paid from a federal vaccine injury fund.[2] Hannah was born in 1999. At 19 months, her pediatrician noted that she was "alert and active" and "spoke well." Then, according to her mother, Hannah received five shots with nine doses of vaccines on a single day, July 19, 2000, and almost immediately developed fever, seizures, and other severe health problems.[3] On October 25, 2002, Terry and Jon Poling filed a petition asking for compensation under the National Vaccine Injury Compensation Program (VICP), blaming the vaccines that Hannah received for her condition. Interestingly, the petition

Trusting Science. Kostas Kampourakis, Oxford University Press. © Oxford University Press (2025).
DOI: 10.1093/oso/9780197787106.003.0009

neither contained a detailed account of Hannah's vaccinations and disorder, nor were her medical records required to accompany it.[4]

The main goal of the VICP is to resolve vaccine injury petitions. It was created by the National Childhood Vaccine Injury Act of 1986 and began on October 1, 1988, after a series of lawsuits threatened to cause vaccine shortages and reduce vaccination rates in the USA. Previous cases with defective vaccines, such as the Cutter incident (Chapter 5), had set a legal precedent: Even if a vaccine-producing company was not found to be liable for negligence, it could still be forced to pay damages for a perceived defective product. Thus, these companies had started stepping away from vaccine production because of the legal procedures taken by families against them. For instance, given the scare that *Vaccine Roulette* had caused (Chapter 6), by 1986, all but one manufacturer of the diphtheria, tetanus, and pertussis (DTP) vaccine had left production. The VICP was founded to solve this problem. The assumption underlying its foundation was that if the government somehow took over the burden of compensation, this would allow the companies to maintain vaccine production. Thereafter, if it was established that vaccines caused an adverse event, children and their families were compensated quickly and generously. As a result, the number of lawsuits against vaccine manufacturers decreased dramatically.[5]

The VICP is administered through the Department of Health and Human Services, which is represented in court by the Department of Justice. The final decision about whether a petitioner should be compensated is made by the US Court of Federal Claims (hereafter the Court). A petition can be filed by, or on behalf of, any individual, of any age, who has received a covered vaccine and believes they were injured as a result. With limited exceptions, all petitions must be filed within three years after the first symptom of the alleged vaccine injury, or within two years of the death and four years after the first symptom of the alleged vaccine injury that resulted in death.[6] Since 1988, over 26,862 petitions have been filed with the VICP. Of these, 22,983 petitions have been adjudicated, with 10,371 of those determined to be compensable, and 12,612 dismissed. The total compensation paid over the life of the program is approximately $5 billion.[7] This money, which is used for compensations, lawyers' fees, and other costs, comes from a trust fund based on a 75-cent excise tax on each vaccine dose sold.[8] It is interesting to note that the VICP website states that "It provides compensation to people found to be injured by certain vaccines. Even in cases in which such a finding is not made, petitioners may receive compensation through a settlement."[9]

This is indeed a means to alleviate pressure from companies manufacturing vaccines. But it may not be all it is. According to legal scholar Anna Kirkland, legal actors can help constitute what she describes as the immunization social order: "... the set of institutions, laws, pharmaceutical biotechnologies, and social practices that work together to produce high levels of vaccine coverage to prevent a wide range of diseases."[10] The cases compensated for vaccine injuries "... are a concession that some people will suffer for the population-level freedom from vaccine-preventable diseases we all enjoy." However, this creates a tension that is not easy to handle: "The vaccine court must do justice to those people—who may still accept the value of vaccines and the immunization social order overall—while managing the confrontations of a hostile social movement eager to exploit this tension."[11]

Back to Hannah Poling. In 2006, her father Jon was the first author of a scientific paper that presented Hannah's case. She was not referred to by name, but as a "19-month-old girl." Here is how her development was described in that article:

> Within 48 hours after immunizations to diphtheria, tetanus, and pertussis; Haemophilus influenzae B; measles, mumps, and rubella; polio; and varicella (Varivax), the patient developed a fever to 38.9°C, inconsolable crying, irritability, and lethargy and refused to walk. Four days later, the patient was waking up multiple times in the night, having episodes of opisthotonos [dramatic abnormal posture due to spastic contraction of the extensor muscles of the neck, trunk, and lower extremities that produces a severe backward arching from neck to heel], and could no longer normally climb stairs. Instead, she crawled up and down the stairs. Low-grade intermittent fever was noted for the next 12 days. Ten days following immunization, the patient developed a generalized erythematous macular rash beginning in the abdomen. The patient's pediatrician diagnosed this as due to varicella vaccination. For 3 months, the patient was irritable and increasingly less responsive verbally, after which the patient's family noted clear autistic behaviors, such as spinning, gaze avoidance, disrupted sleep/wake cycle, and perseveration on specific television programs. All expressive language was lost by 22 months.[12]

We thus see in a scientific paper a clear attribution of this condition to vaccines, without however providing any evidence besides the details of a single individual.

The petition that Hannah's parents submitted demanded that she should be awarded compensation for a vaccine-related disease. The adjudicators for these petitions are called Special Masters and they are judges trained in vaccine issues. Based on the review of the petition and her medical records, Special Master Patricia Campbell Smith decided that the "facts of this case meet the statutory criteria for demonstrating that the vaccination Hannah received on July 19, 2000, significantly aggravated an underlying mitochondrial disorder, which predisposed her to deficits in cellular energy metabolism and manifested as a regressive encephalopathy with features of autism spectrum disorder."[13] We therefore see the main argument: The vaccines are blamed not for directly causing Hannah's disorders, but for worsening an already existing disorder. To support their claim that Hannah's complex partial seizure disorder was related to her vaccine-related injury, her parents also submitted an expert report by Andrew Zimmerman, MD, Hannah's treating neurologist (who was coauthor with Hannah's father in the aforementioned 2006 scientific paper about her). Hannah's parents also requested "an Order permitting the parties, or their representatives, to freely discuss with any person each and every aspect of this case, including the details of the Respondent's concession that Hannah is entitled to compensation for her vaccine-related injuries, including her autism."[14] Eventually, Hannah was awarded a compensation for her vaccine-related injuries. As Special Master Patricia Campbell Smith wrote in a 2011 document: "Based on the persuasive factors supporting petitioner's vaccine claim and respondent's election not to challenge petitioner's claim, the undersigned issued a decision finding that petitioner is entitled to compensation under the Vaccine Program on July 21, 2010, and awarding damages."[15]

What was this compensation? You will not find a decision on the webpage of the United States Court of Federal Claims, referring to petitioners Hannah Poling, a minor represented by her parents and natural guardians, Terry Poling and Jon Poling vs. the Secretary of Health and Human Services. However, the 2011 document by Master Campbell Smith mentions another document, and the dates therein correspond to this decision: Child Doe/77 v. Sec'y of Health & Human Servs., 2010 WL 3395654 (Fed. Cl. July 21, 2010). In that document, reference is made instead to "CHILD DOE/77, a minor," which was represented by "JANE DOE/77 AND JOHN DOE/77" as parents and natural guardians. According to this decision, the petitioners were awarded: "A lump sum payment of $1,507,284.67, representing compensation for life care expenses expected to be incurred during the first year after judgment ($624,713.32), lost future earnings ($674,410.67) and pain and suffering ($208,160.68), in the form of a check payable to petitioners, as the court appointed guardian(s)/conservator(s) of the estate of Child Doe/77, for the benefit of Child Doe/77." An additional payment of $140,109.67 was awarded, "representing compensation for past unreimbursable expenses, payable to John and Jane Doe/77, petitioners."[16]

But why did this decision not refer to the Polings by their real names? In the previous chapter, we saw that there were arguments that thimerosal causes autism. A large legal action thus emerged, known as the Autism Omnibus.[17] Remember that the VCIP was created due to concerns that vaccine manufacturers would abandon vaccine production due to liability concerns, and so its role was to compensate for vaccine-attributed injuries. Given the number of cases, the Special Masters decided to focus on a small number of "test cases" in order to test the claim upon which the petitions for compensation were based: that some combination of mercury and vaccines triggered autism in these children. The decisions about these cases would provide precedents against which future cases would be ruled. After the test cases were completed, specific instructions were given to petitioners as to how to proceed.[18] It seems that the Hannah Poling case was initially one of these test cases, but for some unknown reason it was removed and settled separately. A journalist named David Kirby wrote about the government concession and seems to have leaked its entire text.[19] The vaccine-skeptic website Health Choice Vermont has posted an article under the title "Who was Child Doe 77?," disclosing that it was Hannah Poling and suggesting that "Because Child Doe 77 was removed as a test case, it could not be used to establish precedent on any of the other OAP cases. However, it seems there was misconduct at play."[20] In other words, it was implied that because the Hannah Poling case resulted in a financial compensation, and thus was a defeat for vaccine proponents, it was decided that it be settled separately so that it would not be included among the other test cases and that it could not be used as a legal precedent for future cases.

There were different reactions to the news about the settlement and the more than 1.5 million dollars in compensation. John Gilmore, executive director of the group Autism United, stated that "This decision gives people significant reason to be cautious about vaccinating their children," having filed his own claim about his son developing autism because of vaccinations. According to Hannah's mother, Terry, there were two "theories" about what had happened to Hannah. The first was that she had an underlying mitochondrial disorder aggravated by vaccinations; the second was that vaccinations caused her disorder. "The government chose to believe the first theory," Ms. Poling said, adding that "We don't know that she had an underlying disorder." Hannah's father commented about vaccines that "They're not safe for everybody, and one person for whom they proved unsafe happened to be my daughter." However, experts disagreed. Dr. Edwin Trevathan, director of the National Center for Birth Defects and Developmental Disabilities, said: "I don't think we have any science that would lead us to believe that mitochondrial disorders are caused by vaccines." Dr. Julie Gerberding, director of the Centers for Disease Control, stated: "Let me be very clear that the government has made absolutely no statement indicating that vaccines are a cause of autism. That is a complete mischaracterization of the findings of the case and a complete mischaracterization of any of the science that we have at our disposal today."[21]

David Kirby thought otherwise. As he wrote in June 2008, "When news of the two Poling concessions began to emerge in March, officials from the CDC and other agencies were quick to mount a defensive public relations posture, one that still clouds and confuses the importance of this seminal case, but now seems to be lifting like the haze over Atlanta." The reason for this was the announcement that "On June 29, HHS, CDC, FDA and NIH will hold a major public workshop on mitochondrial disorders, autism and 'triggers for neurological deterioration.'" In his view, it was Hannah's case that made these governmental bodies reconsider what was known about vaccines. As the title of his essay put it, "Hannah Poling really did change everything."[22]

Does concession equal admission of wrongdoing?

Many have argued that the successful claims under the VICP, as in Hannah's case, are clear examples of concessions by the government that vaccines cause autism. And there is no question that this argument makes sense. If vaccines do not cause autism, then why did the VICP compensate Hannah's family? In the case of Porter Bridges, who received compensation from the VICP, his mother Sarah stated in 2003: "… we summarized all of his medical records as we filed his vaccine injury with the special governmental court in charge of these cases. Despite the interventions, his seizures were intractable and his other disabilities unchanged. Results from his special education testing confirmed what we'd suspected for a long time—Porter was severely retarded. The government received our petition and immediately conceded. A vaccine injury had caused permanent brain damage."[23] Sarah confirmed that she had received a compensation of $20 million, with $800,000 to compensate Porter for a lifetime of lost wages and the rest for round-the-clock care for him.[24] If someone files a petition to the VICP and receives compensation, it is reasonable to think that the government has conceded that vaccines cause autism and therefore that there are official concerns about vaccine safety, isn't it?

However, this is not at all how the VICP perceives the situation. Here is how they answer the relevant question on their website:

Do the National Vaccine Injury Compensation Program (VICP) settlements indicate any safety concerns by the Department of Health and Human Services regarding the vaccine alleged to cause the injury?

Conclusions regarding vaccine safety should not be drawn from the fact that cases were settled. Settlements are one way of quickly resolving a petition.

Settlements are an agreement between the respondent (the U.S. Department of Health and Human Services, represented by the U.S. Department of Justice) and the petitioner (the person who filed the vaccine injury petition).

Settlements are not an admission by the United States or the Secretary of Health and Human Services that the vaccine caused the petitioner's alleged injuries.

In settled cases, the United States Court of Federal Claims does not determine that the vaccine caused the injury. Petitions may be resolved by settlement for many reasons, including:

- consideration of prior court decisions;
- a recognition by both parties that there is a risk of loss in proceeding to a decision by the Court making the certainty of settlement more desirable;
- a desire by both parties to minimize the time and expense associated with litigating a case to conclusion; and/or
- a desire by both parties to resolve a case quickly and efficiently.[25]

What is required is that there is a preponderance of evidence, in other words that the vaccine made the injury more likely to have occurred than not. This simply means that the person evaluating the evidence believes that it is more likely than 50% for the claim to be true. This entails that the decisions of the VICP are based on a legal standard that is lower than the standards required for most scientists to think that a conclusion is sufficiently supported. It may be worth mentioning here that this standard is significantly lower than the standard for criminal conviction. It may also be worth pointing out that it seems lower than what we typically think is required to make a belief rational, for example, tossing a coin with an extremely slight bias (makes heads 0.5000001 likely) would satisfy this standard, but it doesn't seem to make rational the belief that the coin will land heads before it is tossed.

A recognition of vaccine injury is made either when the condition is added to the Vaccine Injury Table (a reference table that includes every compensable vaccination and injury in the VICP) or when an individual petition is considered compensable. However, the compensated claims in no way provide an uncontested record of recognition. One reason is that many of the early

DTP compensations were made on dubious grounds. Another is that the compensations often include cases where the vaccine aggravated a preexisting condition, and so it cannot apply in all other cases. The most important issue is that the accounts given in support of the petitioners by treating physicians, who describe how a condition emerged after vaccination, may leave out other possible causes and are also difficult to be countered on the basis of epidemiological evidence, or lack thereof, about a link between the particular vaccine and the particular condition. Because these cases of injury are so rare, petitioners can even argue that it is not possible to have the epidemiological evidence to establish this link. This is why the Court does not require petitioners to provide epidemiological evidence and why the law dictates that the proposed explanation is just legally probable, not scientifically certain.[26] This superposition of legal reasoning over scientific reasoning may have made people distrust the latter.

In a 2008 commentary, soon after the Poling case became publicly known, pediatrician and vaccine expert Paul Offit commented that the VICP "seems to have turned its back on science." The reason for this was that since 2005, the VICP has ruled that if a petitioner proposed a biologically plausible mechanism by which a vaccine could cause harm, as well as a logical sequence of cause and effect, a compensation should be granted. Whether or not solid scientific evidence existed to support such a claim did not make any difference. With respect to the Poling case, Offit noted that although it could be reasonably argued that the development of fever and a varicella vaccine rash after the administration of nine vaccines was sufficient to stress a child with mitochondrial enzyme deficiency, Hannah also had other immunologic challenges that were not related to vaccines. Offit's suggestion was that the VICP more rigorously defined the criteria by which it was determined whether a vaccine had caused harm. "Otherwise, the message that the program inadvertently sends to the public will further erode confidence in vaccines and hurt those whom it is charged with protecting."[27]

A close study of the various cases shows that in all its years of existence, the VICP has never compensated a child on the theory that vaccines caused that child's autism. In contrast, such claims have always been rejected in detailed, well-reasoned decisions in the Omnibus Autism Proceeding. Only when taken out of their context and with their actual content ignored can the VICP cases be used to support the argument that vaccines cause autism.[28] However, for those who are not familiar with the details, it does seem as if the political decision about the creation of the VICP, and the compensations delivered by it, stand in conflict with the view that vaccines are safe and effective. If they were, then there would be no need for a VICP or any kind of compensation, many people claim. The key issue in the Hannah Poling case was that the government had administered a medical product that caused injury to a child. It was not the available scientific evidence that made some people believe this, but a juridical decision. That the evidential basis for this decision was not as solid as scientific rigor demands, and in fact not at all in agreement with the scientific evidence, did not matter. The political decision to create the VICP and the juridical decision to compensate Hannah, and actually do so separately from the other test cases, could thus have been a cause of distrust in science.

Another political decision causing distrust was the one to take thimerosal out of childhood vaccines in the United States in 2001.

Does removing mercury from vaccines mean it is harmful?

In Chapter 8, we saw Burton's campaign against mercury in vaccines, even after thimerosal was removed from all but one vaccine in the USA. He was not alone in this. Several celebrities have also publicly criticized vaccines.[29] For instance, in 2015, actor Jim Carrey severely criticized the Governor of California Jerry Brown for signing a law that barred religious and other personal-belief exemptions for schoolchildren.[30] Carrey tweeted: "California Gov says yes to poisoning more children with mercury and aluminum in mandatory vaccines. This corporate fascist must be stopped." This he said, despite the fact that thimerosal had long been removed from most vaccines. Carrey also tweeted: "I am not anti-vaccine. I am anti-thimerosal, anti-mercury. They have taken some of the mercury laden thimerosal out of vaccines. NOT ALL!"[31] What we see here is a citizen questioning whether all thimerosal has been removed from vaccines. There seems to be no doubt in his mind that thimerosal is dangerous. As one might legitimately ask: If it was not dangerous, then why was it removed?

As we saw in Chapter 8, thimerosal is a mercury-based preservative that has been used for decades in the USA in multidose vials of vaccines. There is no evidence of harm caused by the low doses of thimerosal in vaccines, except for minor reactions like redness and swelling at the injection site. However, in July 1999, the Public Health Service agencies, the American Academy of Pediatrics, and vaccine manufacturers decided that thimerosal should be reduced or eliminated in vaccines as a precautionary measure.[32] The announcement stated the following:

> There is a significant safety margin incorporated into all the acceptable mercury exposure limits. Furthermore, there are no data or evidence of any harm caused

by the level of exposure that some children may have encountered in following the existing immunization schedule. Infants and children who have received thimerosal-containing vaccines do not need to be tested for mercury exposure.

The recognition that some children could be exposed to a cumulative level of mercury over the first 6 months of life that exceeds one of the federal guidelines on methyl mercury now requires a weighing of two different types of risks when vaccinating infants. On the one hand, there is the known serious risk of diseases and deaths caused by failure to immunize our infants against vaccine-preventable infectious diseases; on the other, there is the unknown and probably much smaller risk, if any, of neurodevelopmental effects posed by exposure to thimerosal. The large risks of not vaccinating children far outweigh the unknown and probably much smaller risk, if any, of cumulative exposure to thimerosal-containing vaccines over the first 6 months of life.

Nevertheless, because any potential risk is of concern, the Public Health Service (PHS), the American Academy of Pediatrics (AAP), and vaccine manufacturers agree that thimerosal-containing vaccines should be removed as soon as possible. Similar conclusions were reached this year in a meeting attended by European regulatory agencies, European vaccine manufacturers, and FDA, which examined the use of thimerosal-containing vaccines produced or sold in European countries.

PHS and AAP are working collaboratively to assure that the replacement of thimerosal-containing vaccines takes place as expeditiously as possible while at the same time ensuring that our high vaccination coverage levels and their associated low disease levels throughout our entire childhood population are maintained.[33]

This is a bizarre statement. We are initially told that there is no evidence that mercury in vaccines could have caused any harm to children, as well as that even if such a risk existed, it would be far outweighed by the risk of developing the disease against which the vaccination is done. "Nevertheless, because any potential risk is of concern," it was decided "that thimerosal-containing vaccines should be removed as soon as possible." But why remove thimerosal if there is no evidence of harm?

This is a case of the application of what is called the "precautionary principle." When it became clear that industrialization and the growing human population affected the natural environment, it was decided that measures should be taken in order to repair the damages inflicted upon it. A first step was to require polluters to pay for the cost of the pollution they caused (the *Polluter Pays Principle*). However, it was soon realized that this was not enough, as it also required a kind of preventive policy that would limit the potential damage. It was thus thought that it would be better to follow a "prevention is better than cure" model, according to which we could rely on science to reliably assess and quantify risks in order to limit or diminish any possible damage (*Prevention Principle*). Unfortunately, neither that proved to be sufficient, because of the emergence of unpredictable, uncertain, unquantifiable, and possibly catastrophic risks such as those associated with climate change. This brought about a third, anticipatory model—the *Precautionary Principle* (PP): Whenever human activities might lead to morally unacceptable harm that is scientifically plausible but uncertain, interventions should be taken to avoid or diminish that harm.[34]

This is exactly what happened with thimerosal. The key figure in this story was Neal Halsey, a pediatrician from Johns Hopkins Medical School. In June 1999, he was invited to a Food and Drug Administration (FDA) meeting where it was reported that the amount of ethylmercury contained in vaccines had exceeded the amount that the Environmental Protection Agency (EPA) considered safe for methylmercury. It is worth reminding here that high exposure levels of methylmercury can be toxic, whereas ethylmercury is quickly eliminated and thus not expected to accumulate in the body and cause harm. Halsey was nevertheless shocked not only because he considered the levels of ethylmercury to be high but also because nobody had bothered to look into this before. Not having studies comparing children who had received vaccines with thimerosal to those who had not, Halsey campaigned for refraining from giving to children any vaccine that contained thimerosal. He thus convinced the American Academy of Pediatrics and the Public Health Service to make the statement quoted above, with which they urged vaccine manufacturers to remove thimerosal as quickly as possible and advised pediatricians to postpone giving most newborns the birth dose of the hepatitis B vaccine.[35]

In an editorial written for a prestigious scientific journal a few months later, Halsey clarified his view:

> Exposure to ethylmercury from vaccines containing thimerosal in the first 6 months of life ranges from 0 to 187 µg based on which vaccines are administered. Since many vaccines do not contain thimerosal, most children receive less than the total amount of mercury indicated in the guidelines during the first 6 months of life. If all thimerosal-containing vaccines are given, the total exposures exceed the EPA guidelines, and possibly other guidelines, for the smallest infants. There are safety or uncertainty factors (10-fold for the EPA) built into the guidelines, and experts believe there is no evidence of harm from exposure to thimerosal in vaccines. However, clinicians are uncertain as to how much mercury can be safely given at 1 time when multiple thimerosal-containing vaccines are administered simultaneously.[36]

This is important to note. Halsey was clear that there was no evidence that thimerosal did harm. But there was also uncertainty about how much thimerosal could be tolerated by infants. He concluded his editorial by noting that "Further reductions or elimination of mercury in vaccines will help maintain public confidence by demonstrating a commitment to provide the safest vaccines possible for protecting children against disease."[37]

But in spite of Halsey's honest concerns, the decision to remove thimerosal did not increase the confidence in vaccines; it rather backfired. As journalist Arthur Allen put it in *The New York Times*, "... taking the precautionary step of eliminating thimerosal would be read as an admission of fault."[38] Halsey became the hero of vaccine activists who filed lawsuits against vaccine manufacturers, arguing that thimerosal had caused autism. An expression of concern and a measure of precaution without any evidence of harm, but only due to uncertainty, was interpreted as an admission of years of wrongdoing. Removing thimerosal was a political decision that was not made on the basis of evidence that thimerosal was, or might be, harmful, but due to the lack thereof. The decision was however interpreted as a post-damage measure due to evidence that harm had been done, not as a precautionary one due to uncertainty.

Responding to a criticism by colleagues in 2001, Halsey wrote:

> To maintain public confidence in vaccines, we must ensure the public that safety is taken very seriously, and, when indicated, timely actions are taken to reduce potential risks. Vaccine manufacturers and the FDA should be applauded for the rapid changes in manufacturing and marketing practices that led to an elimination of the use of thimerosal as a preservative in routine vaccines for infants.[39]

Not unreasonable, in principle. But it also depends on the context. Paul Offit has argued that "... it is virtually impossible to exercise the precautionary principle in the United States without significant collateral damage."[40]

In 2004, an expert report was published by the National Academies of Science, titled "Immunization Safety Review: Vaccines and Autism." Its goal was to investigate whether vaccines, in particular the measles, mumps, and rubella (MMR) vaccine and the thimerosal-containing vaccines, were "causally associated with autism." After reviewing the available epidemiological studies about the possible causal link, as well as the studies of the biological mechanisms by which vaccines might have caused autism, the committee concluded that the "... body of epidemiological evidence favors rejection of a causal relationship between the MMR vaccine and autism. The committee also concludes that the body of epidemiological evidence favors rejection of a causal relationship between thimerosal-containing vaccines and autism. The committee further finds that potential biological mechanisms for vaccine-induced autism that have been generated to date are theoretical only."[41] A political decision about a measure had no real basis on the available scientific evidence.

It will be no surprise to find out that political decisions about measures not in alignment with the current scientific knowledge have been taken in other domains besides vaccination. But the harm that such a misalignment has done has never been as significant as in the case of eugenics.

Was eugenics legislation based on solid science?

As briefly mentioned in Chapter 4, the key idea behind eugenics was that inherent biological differences existed among individuals as well as among social groups. Its main goals were the betterment of society by promoting the breeding among "superior" people, while limiting the breeding among "inferior" people. The foundations of eugenics were set by Victorian polymath Francis Galton, who first expressed such ideas in his 1865 paper "Hereditary Talent and Character," wondering how much the offspring would be improved if distinguished women were commonly married to distinguished men.[42] But it was only in 1883 that Galton gave this project the name "eugenics." He coined the term from the Greek word "eugenes," meaning "good in stock, hereditary endowed with noble qualities."[43] Galton published various mathematical analyses of human trait variation related to his eugenics aspirations (he was more interested in promoting the reproduction of those with good qualities than limiting the reproduction of less good ones). However, the movement really went further due to Karl Pearson, who was also instrumental in developing key statistical methods for the analysis of data. He worked with Galton for several years, and upon the latter's death in 1911, Pearson became Galton Professor of Eugenics at University College London. This gave him ample time to work on eugenics projects, always from a mathematical perspective like Galton. He also considered variation as continuous, that is, things were not either black or white but could be either of these or one of the various shades of gray in between.

While all this was happening, zoologist William Bateson managed to establish an influential research program based on a reinterpretation of an 1866 paper by a friar called Gregor Mendel (who was not the father of genetics in the modern sense, in contrast to what you have probably learned in school). As a result, an overtly simplistic perception of heredity prevailed, which roughly considered various traits as "controlled" by specific factors (called "genes" after 1909). This simplistic perception was not problematic for animals and plants studied under laboratory conditions. When the environmental conditions are strictly controlled, as in a laboratory, then one can look for possible causal relations between traits and genes because it is only the latter that can make a difference—all else being equal. Simplifications and idealizations are accepted in research when one is trying to study a phenomenon. The problem however in this case was that the simple model that worked for laboratory organisms and tried to establish the effect of genes when the environment did not change was extrapolated to complex human traits under the implicit assumption that the environment did not matter and that these complex traits were inheritable.[44]

The person who was instrumental for the extrapolation of simplistic genetic models to the inheritance of complex human traits was Charles Davenport. Whereas he had met Galton and Pearson in the 1890s and had been influenced by them, he was not initially interested in eugenics. But once Mendelian genetics became the norm in research, he changed his views. Having persuaded the Carnegie Institution to fund a station for the experimental study of evolution in 1904 at Cold Spring Harbor, with himself as its director, as time went by, Davenport became more and more interested in human traits. He decided to study human heredity by accumulating systematic and detailed data from human pedigrees. His quest to find support for this endeavor resulted in the foundation of the Eugenics Record Office in 1910, with Harry Laughlin as its superintendent. But the most crucial interaction initially was the one that Davenport had with Henry Goddard, Director of Research at the Vineland Training School for the Feeble-Minded in New Jersey. Their exchanges had as a result the popularization of the idea that the condition they described as "feeble-mindedness" (which was never defined precisely, and which was interpreted for convenience) was inherited as a simple recessive trait, that is, one that can be "hidden" in an individual but transmitted and expressed in their offspring.

This idea was popularized through Goddard's 1912 book, *The Kallikak Family: A Study in the Heredity of Feeble-Mindedness*. That was a family with a "good" and a "bad" line, both started by Martin Kallikak after the Revolutionary War. In 1776, he met a "feeble-minded" girl in a tavern and fathered her "feeble-minded" son, Martin Kallikak Jr. Goddard found that among the latter's 480 descendants, 143 were feeble-minded, whereas only 46 were normal, the rest being unknown or doubtful. Then in 1779, Martin married a young lady from a good family, from where the "good" family line emerged. Among 496 of its members, only 3 had been found to be "somewhat degenerate" but not defective. For Goddard, the proportion of "feeble-minded" people in the "bad" line and their total absence in the "good" line of the Kallikak family was sufficient proof that "feeble-mindedness" was inherited like other physical features and could not be changed by interventions in the environment, such as by educating those people.[45] Initially, and for some time, many geneticists accepted Goddard's conclusions as a fact.[46]

Laughlin was among those convinced. He was an ardent eugenicist who was eager to take legal measures that would improve society. According to eugenicist views, society could only be improved if these defective, "feeble-minded," people did not have any offspring. This could be achieved in two ways: either with segregation in institutions so that "feeble-minded" people could not meet and mate or with sterilization so that they could have no offspring at all. The first sterilization laws in the United States had been introduced in Indiana in 1907 and in Washington and California in 1909. However, except for California, they were not really implemented. By 1921, 2,248 sterilizations had been performed in California, which represented more than 80% of all cases in the United States.[47] Laughlin thought that something had to be done about this. He worked for years to bring together in a single book all the knowledge about sterilization, but had trouble finding a publisher for it. Eventually, it was Harry Olson, chief judge of the Municipal Court of Chicago and advocate of eugenic sterilization, who helped Laughlin get the book published in 1922. It contained not only the history of the legal movement for sterilization in the United States but also a Model Law that Laughlin had developed some years ago and all the relevant "scientific facts."

In Chapter X, titled "The Right of the State to Limit Human Reproduction in the Interests of Race Betterment," Laughlin defended sterilization as an act that contributed to the common good. His argument was that states often impose measures that may be fatal for individuals, if these serve the good of the broader society, giving military conscription as an example. "It is absolutely non-punitive, but is demanded by the principle that, in the long run, the welfare of the commonwealth is of vastly more importance to the sum total of human happiness than is the temporary freedom and personal security of the individual."[48] But the most striking analogy Laughlin made was, in my view, that with compulsory vaccination:

Compulsory Vaccination is analogous to compulsory eugenical sterilization to the extent that both are non-punitive and that both involve the seizure of the individual and subjecting him or her to surgical treatment. Both vaccination and sterilization are done supposedly for the public good. Vaccination protects the individual and his associates from a serious and loathsome disease in the more immediate future; eugenical sterilization protects society from racial degeneracy in the more remote future.[49]

It does not take a lot of thought to realize that this analogy is simply wrong. Whereas there was at the time plenty of evidence that vaccines can protect from infectious diseases, there was no real evidence that sterilization would confer any advantage to society. Whatever caused the degeneration of society, such as "feeble-mindedness," should be purely innate and therefore inheritable for sterilization to have any positive effect—leaving aside the ethical aspects of such a coercive practice. But the evidence at the time did not support any such view.

One of the first geneticists to openly criticize the hereditary assumptions underlying sterilization policies was Thomas Hunt Morgan, who wrote in 1924:

> In the case of man, I have pointed out that we use the word inheritance in a double sense. In a biological sense it means one thing, in a social sense it means something quite different. While these two aspects of human heredity have seldom been confused by those writing on the subject, nevertheless, I can not but think that our familiarity with the process of social in heritance is responsible, in part, for a widespread inclination to accept uncritically every claim that is advanced as furnishing evidence that bodily and mental changes are also transmitted.[50]

Morgan's conclusion in the second edition of his book *Evolution and Genetics*, published in 1925, was even more caustic:

> If within each human social group the geneticist finds it impossible to discover, with any reasonable certainty, the genetic basis of behavior, the problems must seem extraordinarily difficult when groups are contrasted with each other where the differences are obviously connected not only with material advantages and disadvantages resulting from location, climate, soil, and mineral wealth, but with traditions, customs, religions, taboos, conventions, and prejudices. A little goodwill might seem more fitting in treating these complicated questions than the attitude adopted by some of the modern race-propagandists.[51]

Around the same time, Goddard's theory received a devastating critique by neurologist Abraham Myerson in his 1925 book *The Inheritance of Mental Diseases*. Among his other writings, Myerson critiqued Goddard's 1912 book on the Kallikak family, which had been repeatedly quoted in all the lay literature and which was the key reference for the threat of the predominance of the "feeble-minded." "No royal family has enjoyed quite such & prestige as this group," Myerson noted, with the book itself having "all the dramatic flavor of the missionary spirit, or of one who 'views with alarm' and wishes to awaken into vigilance the threatened normals."[52] Myerson also confessed to "a feeling of shame" for the work done by the fieldworker in that case. Having overseen a clinic to which "feeble-minded" people were brought every day, Myerson was aware that many of those people, who were mur-derers, thieves, sex offenders, and so on, were in the end shown not to be "feeble-minded" at all. In contrast, the mental tests and the psychological examinations had shown that some of those people were "of full average mentality or better."[53] Myerson also questioned the assumptions that the tavern girl with whom Martin Kallikak had an illegitimate "feeble-minded" son, Martin Kallikak Jr., was also "feeble-minded." How is it possible, Myer-son asked, to have this kind of knowledge about a girl who lived so long ago? He compared this to the historical knowledge available for Washing-ton and Lincoln, noting that to acquire this kind of knowledge, historians had devoted their lives to study various kinds of sources. But the same could not be said to be the case either for Martin Kallikak or for the anonymous tavern girl.[54]

But eugenicists were undeterred. Laughlin's book provided the basis for the sterilization law developed in Virginia by Aubrey Strode, a lawyer and a member of the Virginia Senate. In some cases, he even reproduced verbatim Laughlin's arguments. Strode made sure to consider whatever had happened in the past. He also managed to maneuver politically so that the sterilization law was passed. However, given previous lawsuits in Virginia and elsewhere that had challenged the constitutionality of sterilization laws, it was consid-ered necessary to test the constitutionality of this new law with a test (setup) case. The person selected for this test was Carrie Buck who was an inmate in the Virginia State Colony for Epileptics and Feebleminded. Her mother Emma was also institutionalized in the same place for alleged promiscuity, poverty, and perceived mental deficiencies. When this happened, Carrie was a child and was placed in foster care. However, at the age of 17, she became pregnant, allegedly due to rape by a member of her foster family. This resulted in her being institutionalized at the Virginia State Colony for Epileptics and Feebleminded. Because her mother Emma was already there, this was used as evidence for the argument that Carrie came from a "defective" lineage, justi-fying sterilization. Her daughter Vivian was also classified as "feeble-minded." These three generations of "feeble-minded" women seemed like a perfect case.

The juridical case became known as Buck v. Bell, with John Hendren Bell being the superintendent of the Virginia State Colony for Epileptics and Feebleminded at the time. Laughlin testified as an expert providing the supposed evidence in favor of the sterilization law:

> Let me say, also, that in the archives of the Eugenics Record Office there are many hundreds of manuscript pedigrees of families with feebleminded members. These pedigrees prove conclusively that both feeblemindedness and other intelligence levels are, in most cases, accounted for by hereditary qualities.[55]

Eventually, the case reached the Supreme Court of the United States. There was an 8–1 ruling that Carrie should be sterilized, with the famous statement by Justice Oliver Wendell Holmes in 1927:

> It is better for all the world if, instead of waiting to execute degenerate offspring for crime or to let them starve for their imbecility, society can prevent those who are manifestly unfit from continuing their kind. The principle that sustains compulsory vaccination is broad enough to cover cutting the Fallopian tubes ... Three generations of imbeciles are enough.[56]

A consequence of this decision was the legitimization of sterilization laws in the United States. Whereas several states already had sterilization laws, their application was inconsistent and their effects practically nonexistent, perhaps except for California. After this case, many states updated and put into effect their sterilization legislation. In the end, around 60,000–70,000 people were sterilized in the USA during the whole twentieth century. The decision to pass and implement sterilization laws after 1927 was a political decision that was no longer supported by mainstream science.[57] But they were implemented, doing harm.

What is crucial to understand? (and also teach in schools!)

None of the decisions to establish the VICP, to remove thimerosal from vaccines, or to pass sterilization laws, all of them political decisions, were made because there was solid scientific evidence that the injuries of the compensated children were due to vaccines, that thimerosal was dangerous, or that "feeble—mindedness" was inheritable. It is therefore crucial to make the distinction between such political decisions and the relevant science.

That political decisions can be informed by a scientific field does not entail anything for that field itself. The reason is that science is not the only factor upon which political decisions are made. There also exist other important social factors that decision-makers have to consider. A clear example of this was the political decisions about the different kinds of measures taken during the COVID-19 pandemic. These included school and workplace closing, stay-at-home orders, restrictions on gatherings, cancellations of public events, mask orders, restrictions on internal or international travel, testing policies, vaccination policies, contact tracing, and more. There were two main issues with these measures: (1) if they were really efficient in stopping the propagation of SARS-CoV-2, and (2) what their side effects were for everyday life, especially for the economy and the psychological well-being of people. We all know that these measures were bad psychologically speaking, so I will not further consider this issue. However, there were differences in how these measures affected the economy. There were people who kept working from their homes, and others who could not and thus could not earn their living. Because these measures were often recommended by scientific committees, in many cases scientists were blamed for their consequences. But it was government officials who made the decisions, not scientists.

Therefore, a clear distinction must be made between the recommendations of scientists (for instance, for lockdowns during the first months of the COVID-19 pandemic to limit the spread of SARS-CoV-2), and the societal consequences of such measures (for instance, the difficulties for individuals from living a non-normal life in closure, or the financial consequences from not being able to go to work and earn money). What scientists, mostly epidemiologists and virologists, were asked to recommend were ways with which the pandemic could be contained. It is very likely that if in early March 2020 we all stayed inside our homes for two weeks, the pandemic would have been over very soon. But this was entirely impossible to happen for practical reasons; to say the least, people needed to buy food, and someone had to transfer that food wherever it was sold. As a world-level lockdown was not practically possible, scientists had to make other recommendations that limited the transmission of the virus. What scientists had to estimate was which of these measures would be effective: Anything that would limit the propagation of the virus until safe and effective vaccines were made available. The scientists thus made specific recommendations at various stages of the pandemic to this end.

But which of these measures would be implemented, when, and for how long was the decision of governments, not of scientists. These were political decisions informed by science, which nevertheless had to take other factors into account. An important consequence of these restrictive measures was job and income losses. A recent study with 321,000 adults in 117 countries has estimated that about half of the world's adult population lost income during the COVID-19 pandemic. So, whereas there is no doubt that the various restrictions saved many lives until vaccines were made available, they also badly affected many people resulting in financial hardship. The countries that adopted more stringent policies experienced a higher rate of economic harm. Workers with lower income or education were much more likely to report harm linked to the pandemic than those with tertiary education or relatively high incomes, so people were unequally affected. It was also found that the various restriction measures did not have similarly bad consequences: school closings, stay-at-home orders, and other economic restrictions were strongly associated with economic harm, whereas contact tracing and mass testing were not.[58]

You may now wonder: Were scientists not aware that the measures they recommended would have economic side effects? Well, they should have been, and these side effects should have been considered by scientists before making recommendations and by politicians before making decisions. But this is different from blaming scientists for these side effects. Furthermore, we need to be able to look at the broader picture. I understand that choosing between being dead and being in financial hardship is not a real dilemma, but it is a fact that those who did not survive the pandemic could not worry about their financial situation. In Switzerland, where I live, there has been a heated debate about the COVID-19 measures. In fact, it is the only country whose citizens have already voted three times (!) on the legal basis for pandemic measures—and all three times the measures were approved. This was possible because an optional referendum is a constitutional right in Switzerland. When 50,000 signatures from eligible individuals are collected against a law, a referendum is held. The second referendum was the most heated, yet the law was approved by 62% of voters. This of course means that a significant portion of the population did not agree.

However, it is interesting to note that the measures taken in Switzerland, which I can attest were less strict than those taken in other countries, had a positive effect with respect to deaths. Using data for 2011–2019, researchers predicted the expected number of deaths in Switzerland from February 2020 to April 2022 and compared that number with laboratory-confirmed COVID-19 deaths. Overall, they found that COVID-19 was directly responsible for an estimated 18,000 deaths, of which only around 13,000 laboratory-confirmed COVID-19-related deaths were reported. In other words, COVID-19-related mortality was underestimated, as only about

70% of COVID-19-related deaths were confirmed. However, they also found that the observed mortality was 4% less than what was expected and concentrated in the age groups 40–59. The researchers argued that this was probably due to the reductions in mobility, road traffic, air pollution, and sports activities, because reduced mortality was more evident when the measures were most stringent. Despite the many negative effects of the pandemic measures, there were fewer deaths in Switzerland during the pandemic than expected. This entails that the negative effects of restrictive measures were outweighed by the positive effects with respect to mortality.[59]

Vaccination had an even better effect. Even though vaccination against SARS-CoV-2 was never mandatory, there was a point in time after which unvaccinated people could practically do very few things, as they were not allowed in cinemas, restaurants, or other public spaces. One study estimated that from December 1, 2020, to September 30, 2021, COVID-19 vaccination prevented approximately 27 million infections, 1.6 million hospitalizations, and 235,000 deaths in the USA, among vaccinated adults. It was also estimated that for about the same period, vaccination was estimated to prevent more than half of the expected infections, hospitalizations, and deaths in adults.[60] A more recent study, based on data until March 2022, has estimated that COVID-19 vaccination cumulatively averted 2,265,222 deaths and prevented 17,003,960 hospitalizations. The authors noted that until that time, there had been over 4.7 million reported hospitalizations and almost 1 million documented deaths in the USA due to COVID-19, with more than 60% of these deaths occurring after the start of the vaccination program.[61]

Without forgetting the (devastating for some) effects of lockdowns and other restrictive measures on the economy, on education, or on the psychology of most of us, it is important to note that for these to matter, one has to be alive. Therefore, I think it makes sense that reducing mortality was a top priority. Of course, we cannot neglect the other important aspects of life. There is a delicate equilibrium to reach, and we can only hope that whenever the next pandemic comes, we will be able to achieve this equilibrium. Scientifically speaking, scientists' recommendations about restriction measures at the beginning of the pandemic and about the importance of vaccination once vaccines became available later on were correct. But not everything is about science, and not everything is scientists' responsibility. How the measures were implemented, how vaccination was promoted, and whether any of these were enforced by law were not decided by the scientists. It is governments that should be congratulated or blamed. We must distinguish between scientists' recommendations and the respective political decisions.

10

"Bring back our #ChildhoodDiseases they keep you healthy & fight cancer"

The lack of directly observable evidence

Fake hysteria?

> Here we go LOL #measlesoutbreak on #CNN #Fake #Hysteria The entire Baby Boom population alive today had the #Measles as kids Bring back our #ChildhoodDiseases they keep you healthy & fight cancer[1]
> I had the #Measles #Mumps #ChickenPox as a child and so did every kid I knew - Sadly my kids had #MMR so they will never have the life long natural immunity I have. Come breathe on me![2]

Darla Shine's 2019 tweets would likely find many proponents among parents of young children. She is the wife of Bill Shine, a former Fox News president who was the White House Deputy Chief of Staff for Communications when Donald Trump was President of the USA. Her point was that having had measles as a child, like many people of her generation, she now has lifelong natural immunity. But this will not be the case for her children, she notes, who were vaccinated against measles and who now will not have the opportunity to get the natural immunity their mother has. Needless to say, she received criticism on Twitter. She replied:

> People texting I'm spreading lies about #vaccines. I'm retweeting physicians, scientific studies, and questioning why #Media covers #MeaslesOutbreak one-sided. Many of the kids w/ #Measles in #Washington WERE #Vaccinated Go ask their #Governor.[3]

This referred to a 2019 measles outbreak in Clark County, Washington. According to the officials, there were 71 confirmed measles cases, 52 of whom were children younger than 10 years old. Among these 71 people, only 3 had previously received the measles, mumps, and rubella (MMR) vaccine.[4] As we

Trusting Science. Kostas Kampourakis, Oxford University Press. © Oxford University Press (2025). DOI: 10.1093/oso/9780197787106.003.0010

have seen, this is a triple vaccine against measles, mumps, and rubella, which are highly contagious viral diseases for which there is no specific treatment. Vaccination is therefore the only effective means of prevention, and it is recommended that all children between 12 and 24 months of age be vaccinated with two doses of the MMR vaccine.

One of the responses to Shine's tweets is worth quoting: "Hey lady you do realize the reason why YOUR generation does better with health issues isn't because you had MMR it's because you survived it." The same point was made by pediatrician and vaccine expert Paul Offit in an interview: "When Darla Shine talks about how great it was that she had measles as a child, what she forgets to mention is that she gets to tell her story because she's alive. The ones who died—we don't hear from them." Do keep these comments in mind; not all people made it through measles.[5]

In any case, Shine does not represent the majority view in the USA. In a 2023 survey by the Pew Research Center with 10,701 US adults, it was found that about 9 out of 10 of them said that the benefits of the MMR vaccine outweigh the risks. When they were asked to independently assess the health benefits and side effects risk of the MMR vaccine, about 2 out of 3 said that the preventative health benefits of MMR vaccines are either very high or high, and that the risks of side effects are either very low or low. These figures have been pretty much stable since 2016. When asked whether healthy children should be required to be vaccinated in order to attend public schools, 7 out of 10 participants agreed, compared to 8 out of 10 in 2019 and 2016. Furthermore, about 1 in 4 participants suggested that parents should be able to decide not to vaccinate their children, an approximately twofold increase since 2019. This decline in support seems to be related to political orientation, as it is mostly due to the changing views among Republicans, 6 out of 10 of whom now support requiring children to be vaccinated to attend public schools, compared to 8 out of 10 in 2019. In contrast, almost 9 out of 10 Democrats supported school-based vaccine requirements, a view that has not changed significantly since 2016.[6] It should be noted that according to the Centers for Disease Control and Prevention, the USA has maintained a measles elimination status since 2000. This does not mean that cases will cease to exist, but insofar as people are vaccinated, the risk for the majority of the population would be low.[7]

Measles is due to one of the most contagious viruses. One way of measuring the contagiousness of a pathogen is the basic reproduction number (R_0), of which you probably heard in the media during the COVID-19 pandemic. This is the estimated number of new infections for each infected individual. If R_0 is smaller than 1, it means that it is unlikely that an infected person will infect someone else. In such a case, the number of infected individuals in the community is expected to decrease with time. If R_0 is larger than 1, however, it means that each infected individual will infect one or more others. To give some examples, whereas the R_0 of the influenza virus is 1.27–1.8[8] and that of the Delta variant of the SARS-CoV-2 virus is 5–8,[9] the R_0 for the measles virus is usually estimated to be around 12–18 (although it can vary depending on the setting).[10] What this means is that, taking the lowest estimate, after four cycles of transmission, a person infected with the influenza virus will have infected $1.27 \times 1.27 \times 1.27 \times 1.27$, or 1.27^4, that is, between two and three other people; a person infected with the Delta variant of the SARS-CoV-2 virus will have infected 5^4, or 625 other individuals; and a person infected with the measles virus will have infected 12^4, or 20,736 individuals!

The measles virus remains active in the air or on surfaces for up to two hours and spreads through coughing or sneezing, or even breathing. It has been estimated that a person infected with measles can infect 9 out of 10 of their unvaccinated close contacts. Under certain conditions, infection can result in severe complications and deaths. This is why according to the World Health Organization (WHO), countries should aim at achieving a 95% coverage with two vaccine doses. It has been estimated that measles vaccination has prevented 57 million deaths between 2000 and 2022. However, in 2022, there were an estimated 136,000 measles deaths globally, mostly among unvaccinated or undervaccinated children under the age of 5 years. In 2022, about four out of five children all over the world received one dose of the measles vaccine during their first year of life, which is the lowest since 2008.[11]

Figure 10.1 presents the measles cases globally between 1980 and 2022. As is obvious, the cases of measles have been declining continuously, and this is due to vaccination. Still, outbreaks have occurred in various places in the world such as in 2019, when there were almost 1 million cases globally. Henrietta Fore, Executive Director of UNICEF, and Tedros Adhanom Ghebreyesus, Director General of the WHO, described it at the time as a global measles crisis. There were numerous cases all around the world, including countries such as the USA where measles had previously been eliminated. They considered as a main cause the insufficient levels of vaccination, either because children did not have access to vaccines or because there was uncertainty about their necessity and safety. The latter was especially evident in several high- and middle-income countries, such as the United States, Ukraine, France, or the Philippines.[12]

Unfortunately, we have recently started having evidence of what refraining from having our children vaccinated can do. During 2023, more than 58,000 measles cases were reported by 41 of the WHO European region's 53 member

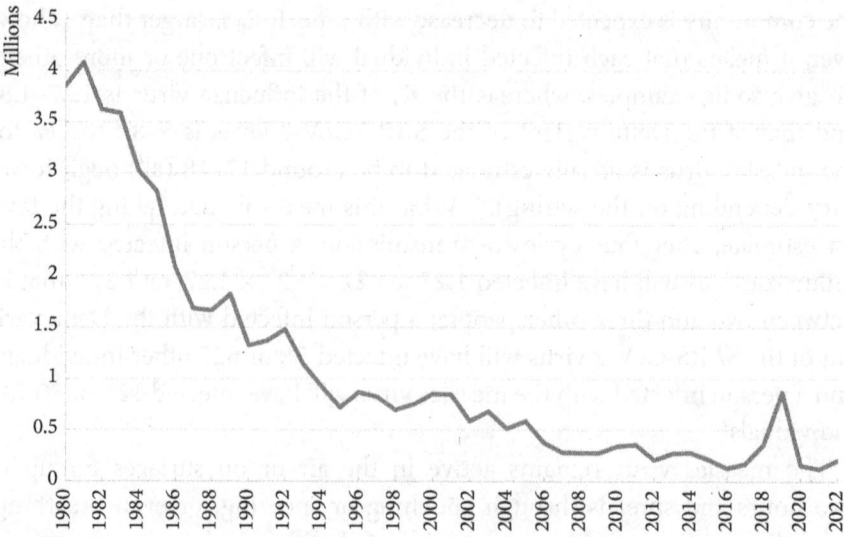

Figure 10.1 Reported worldwide cases of measles, 1980–2022. Based on data from WHO, Global Health Observatory (https://www.who.int/data/gho/data/indicators/indicator-details/GHO/measles-number-of-reported-cases).

states, compared to 941 cases in 2022. The increase in the number of cases resulted in the hospitalization of thousands of people and 10 measles-related deaths.[13] The reason for this, according to the European Centre for Disease Prevention and Control, is that the vaccination levels in Europe are below the required level of 95%, with the average being 89.7% of the population vaccinated with two doses. Only Hungary, Malta, Poland, Portugal, and Slovakia seem to have achieved in 2022 the 95% threshold required for measles.[14] All this entails that contrary to what Shine suggested, there is no fake hysteria about measles.

Seeing for oneself

Why did Shine tweet about a "fake hysteria"? Why did only 7 out of 10 people in the USA consider the preventative health benefits of MMR vaccines as high, when WHO recommends a vaccination coverage of 95%? One important, and often unacknowledged, reason for this is—perhaps ironically—how successful vaccination against measles has been. Since the vaccine was introduced in the 1960s, the cases of measles have dropped significantly. Because of this, some people nowadays think that vaccination against measles can be avoided. Their arguments often take the form "if there is no measles, why should I then give the measles vaccine to my children?"; in short, why undergo a medical intervention for a problem that does not exist? As Offit put it: "Measles is out of sight and out of mind, so we think it's no big deal."[15] Whereas the lack of directly observable evidence of measles makes people question the recommendation to get vaccinated against it, what these people fail to realize is that it is due to the effectiveness of the respective vaccines that measles is rare. So, getting the MMR vaccine is not an unnecessary medical intervention that we make against a problem that does not exist; rather the problem does not exist thanks to that medical intervention, and to keep it that way, vaccination is necessary. This is easily established by the fact that whenever and wherever the vaccination rates have dropped even slightly, measles cases have risen.

The problem therefore is that when people do not see a danger, they think that the danger does not exist—but I noted that it does not exist thanks to vaccines. What makes this even worse is that perhaps these people have no idea what the danger is because they have never encountered cases of particular diseases—again thanks to vaccines. These people therefore believe that there is no reason to insert into the body of their infant substances from a dangerous virus as a precaution against a disease that they do not see anywhere. Whereas they may have no problem giving their sick child a substance against whatever made them sick—because they do see it is sick—they would rather refrain from getting it vaccinated because they do not perceive any danger. But I am inclined to think that if they had seen how devastating some vaccine-preventable diseases are, they would likely change their minds.

Have you seen cases of smallpox (Chapter 3)? Probably not, because smallpox has been eradicated since 1980. Some people, perhaps Darla Shine too, may be sorry for this, as they will not have the opportunity to acquire natural immunity against the virus. But they would be terribly wrong. First, as explained in Chapter 2, vaccines contain molecules of the natural pathogen and therefore cause the same kind of immunity that a natural infection would cause—but of course with no, or much less severe, symptoms. Second, next time someone tells you that vaccines are not necessary, you can show them the photograph in Figure 10.2. This is a photograph taken by Dr. Allan Warner of the Isolation Hospital at Leicester in the UK, which is included in a medical atlas published in the very early twentieth century. Here is the caption: "Shows two boys, both aged 13 years. The one on the right was vaccinated in infancy, the other was not vaccinated. They were both infected from the

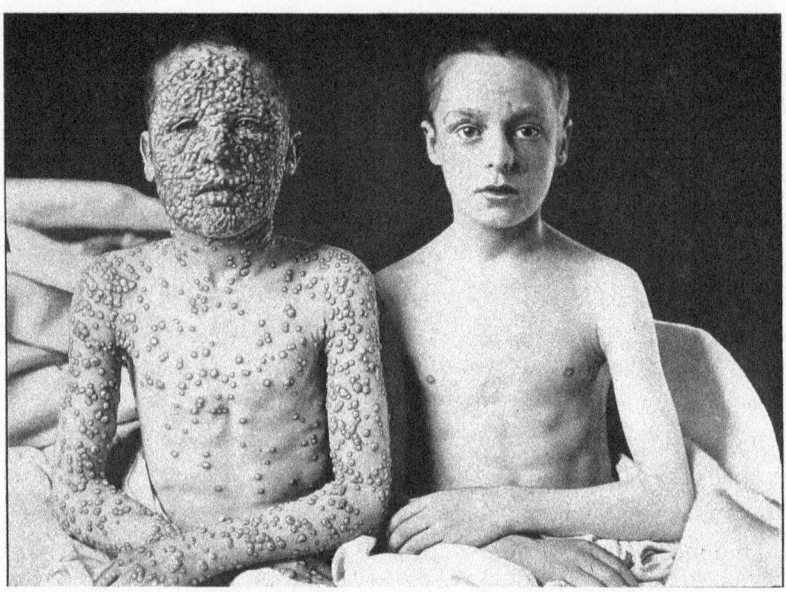

Figure 10.2 Two 13-year-old boys who were both infected by smallpox on the same day from the same source. The boy on the right had been vaccinated in infancy, whereas the boy on the left had not. From Hutchinson, S. J. (Ed) (1901–1903). *Atlas of Illustrations of Clinical Medicine, Surgery and Pathology (a Continuation of the Atlas of Pathology).* London: John Bale, Sons & Danielson Limited, Plate O (Public Domain).

same source on the same day. Note that while the one on the left is in the fully pustular stage, the one on the right has had only one or two spots, which have aborted and have already scabbed."[16] You can thus easily compare smallpox in a vaccinated and an unvaccinated boy. No comments are necessary; this photograph is a perfect instantiation of the adage "A picture is worth a thousand words."

Or consider the photograph in Figure 10.3, where you see two-month-old (!) Martha Ann Murray being watched by nurse Martha Sumner in St. Mary's hospital while she was critically ill of polio (Chapter 5). As we saw, some polio patients were obliged to be inside "iron lungs," with only their head left outside the chamber, in order to be able to breathe as their diaphragm, the muscle that supports our breathing, was paralyzed. Pumps raised and lowered the pressure inside the chamber. When the pressure fell below that of the lungs, outside air filled the lungs; when it rose above that of the lungs, air was expelled from the lungs. The younger people among those who sadly experienced this should be over 70 years old today. This means that most of the people living today have not had a direct experience with the iron lung,

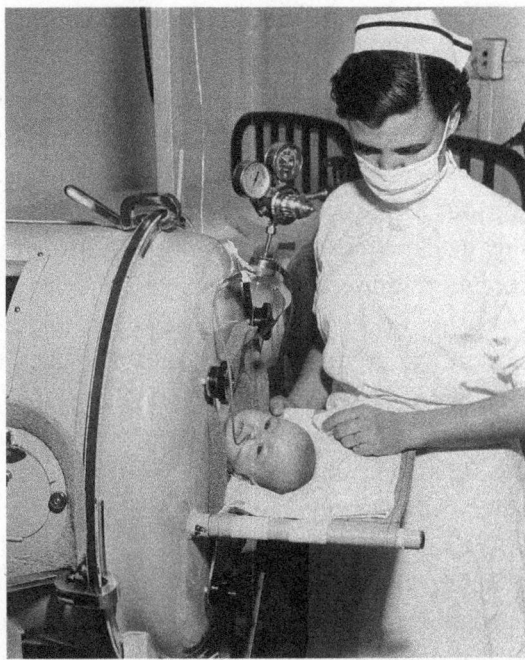

Figure 10.3 Polio strikes two-month-old Martha Ann Murray, who is watched by nurse Martha Sumner in St. Mary's hospital when she is critically ill of polio. Credit: AP Photos.

nor had ever been in the position of Martha Ann Murray's parents who had to watch their baby in this situation.

I am inclined to think that parents who hesitate or refuse to vaccinate their children would think twice after seeing the photographs in Figures 10.2 and 10.3. They clearly show what an infectious disease can do. But the lack of directly observable evidence such as this makes people prone to question the value of vaccines, without realizing that they can do so because they take it for granted. I recall a relative telling me about two other members of our family who were infected by SARS-CoV-2 but nevertheless only had mild symptoms: "You see, it was nothing, just a flu." That these two family members had been fully vaccinated before they were infected did not make any difference to him. Luckily, we did not have to deal with any COVID-related death in the broader family. But many others did, unfortunately.

To better realize the issue of how the lack of directly observable evidence can mislead us, it is worth considering how this occurs in other scientific fields. Evolutionary biology is another scientific discipline in which the lack of directly observable evidence has made people question its premises and conclusions.

The problem of "missing links" in evolution

Evolutionary theory and the idea of biological evolution more broadly have been the topics of fierce public debates. Various polls have shown that the relatively low public acceptance of evolution is related to religious belief, particularly to the various forms of creationism, the idea that a supreme being created organisms in their present form. This is not the only reason that people do not accept evolution, but it is an important one. Details notwithstanding, many who oppose the idea of evolution claim that natural processes cannot alone explain the origin of species and their adaptations. Rather, something more is required. This additional something is usually god, whose existence and power are the ultimate explanation for everything. In this view, whenever there exist incomplete scientific explanations of a natural phenomenon, god's divine intervention stands as the always-sufficient alternative. This idea is described as the "god of the gaps," meaning that whenever there is a "gap" in the explanatory potential of science, this is filled by the assumption that god intervened and made things happen. Therefore, god's intervention stands as the explanation for whatever cannot be explained by science. However, this is not a rational argument. That there is no scientific explanation for something because, say, no relevant evidence has been found yet does not entail that such evidence will never be found.[17]

The classic example here is the quest for transitional fossils, that is, fossils of forms that are intermediate between those of large and well-known groups and that could thus serve as evidence for the evolutionary transition from one group to the other. Fossils are the remains of the hard parts (usually skeletons) of organisms that have been preserved in the form of stone, maintaining the organisms' original shape. For instance, in order for evolutionary biologists to establish the evolution of land organisms from marine ones, there should exist fossil forms that have characteristics intermediate between the two. The lack of existence of such a form can make people think that such a transition is not possible, not least because they cannot really see how it could have been possible. Some may thus be prone to accept explanations of the special creation of marine and land animals. Others may simply question evolutionary theory just because they require evidence that such a transition has indeed been possible, and cannot just blindly accept it. This is not always the case, but as it happens, evidence of this kind is occasionally available.

Even though the evolution of tetrapods (four-limbed vertebrates) from sarcopterygian (lobe-finned) fish was generally accepted by evolutionary biologists, there existed few fossils that might suggest how this evolutionary transition could have taken place. Therefore, it was difficult even for proponents of evolution to explain how this transition could have actually happened. This all changed with the discovery of *Tiktaalik*, the skeleton of which represents a shift from the structure of primitive sarcopterygian fish towards the structure of tetrapods. Paleontologist Neil Shubin and his colleagues were aware that amphibian fossils had been recovered from rocks about 365 million years old and that fish fossils had been recovered from rocks about 385 million years old. Consequently, they knew that they should look for transitional forms in rocks aged 365–385 million years. In addition, knowing that sedimentary rocks usually preserve fossils, they knew they had to look for rocks formed in oceans, lakes, or streams, ruling out volcanic and metamorphic rocks in which fish fossils would not likely be found. Last but not least, they had to choose areas that were not inhabited and where fossils might be exposed on the surface of rocks. They thus decided to look for transitional forms between fish and tetrapods at the Canadian Arctic, where the rocks were of the right age, type, and exposure. And it was there, at the Fram Formation in Nunavut Territory, Canada, where *Tiktaalik* was eventually found.[18] In *Tiktaalik*, which has the best preserved and most fully described pectoral fin skeleton, the shapes of the joint surfaces between the distal elements suggest a capacity for hyperextension, likely associated with upwards flexion of the fin as it was pressed against the substrate by the weight of the body. These characteristics strongly suggest that the pectoral fins were adapted "for walking," with endoskeletons long enough to lift the body off the substrate. In other words, this is a clearly transitional form (Figure 10.4).[19]

Once people see transitional forms like *Tiktaalik*, they might be more willing to accept the fact of evolution, because they can see how it could have occurred.

What is crucial to understand? (and also teach in schools!)

Understanding that science often draws on indirect evidence is key to understanding how it is done. I think that this is best illustrated by considering Abraham Wald's story about the missing data regarding the returning aircraft in World War II. Wald was a Hungarian mathematician, who made important contributions to statistics. During World War II, he was a member of the Statistical Research Group (SRG), a team of mathematicians and statisticians who assisted the US military in solving complex problems. One of these was where to enhance the armor of aircraft to ensure their protection without compromising their maneuverability and fuel efficiency because of

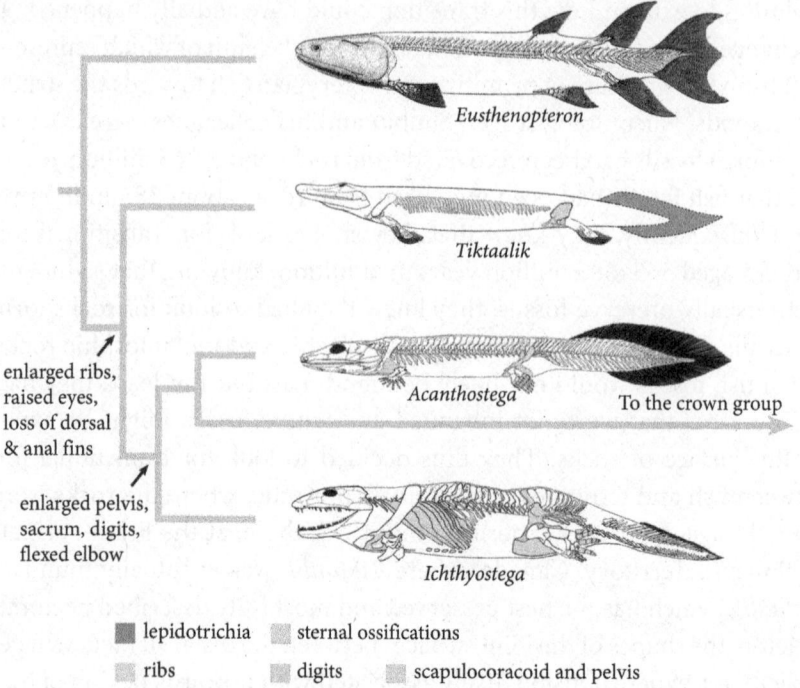

enlarged ribs,
raised eyes,
loss of dorsal
& anal fins

enlarged pelvis,
sacrum, digits,
flexed elbow

Eusthenopteron

Tiktaalik

Acanthostega   To the crown group

Ichthyostega

▓ lepidotrichia   ▓ sternal ossifications

▓ ribs   ▓ digits   ▓ scapulocoracoid and pelvis

Figure 10.4 A phylogenetic tree representing the fish–tetrapod evolutionary transition and illustrating the major anatomical changes. The taxa are arranged from most aquatically adapted at the top to most terrestrially adapted at the bottom. Note the enormous enlargement of the ribcage and pelvis. Ribs are present in *Eusthenopteron*, but so small that they are difficult to see. In the color coding, blue denotes an aquatic adaptation, and yellow to red colors represent adaptations for walking and weight support. It is evident that *Tiktaalik* has features intermediate between those of fish (*Eusthenopteron*) and tetrapods (*Icthyostega*). Reproduced from Ahlberg, P. E. (2018). Follow the footprints and mind the gaps: A new look at the origin of tetrapods. *Earth and Environmental Science Transactions of the Royal Society of Edinburgh*, *109*(1–2), 115–137, under the terms of the Creative Commons Attribution license.

excess weight. Thus, the military approached the SRG with data from operations in Europe. When they analyzed those data, it became evident that the bullet holes on the returning US planes were concentrated in certain areas, particularly the fuselage, whereas there were not that many bullet holes in the cockpit and the engines (Figure 10.5). The military officers had thus arrived at the conclusion that they had better put the additional armor where the most damage had been found: the fuselage of the aircraft. But as they could not decide how much additional armor they should put or on which parts of the fuselage, they approached Wald to ask for his advice.

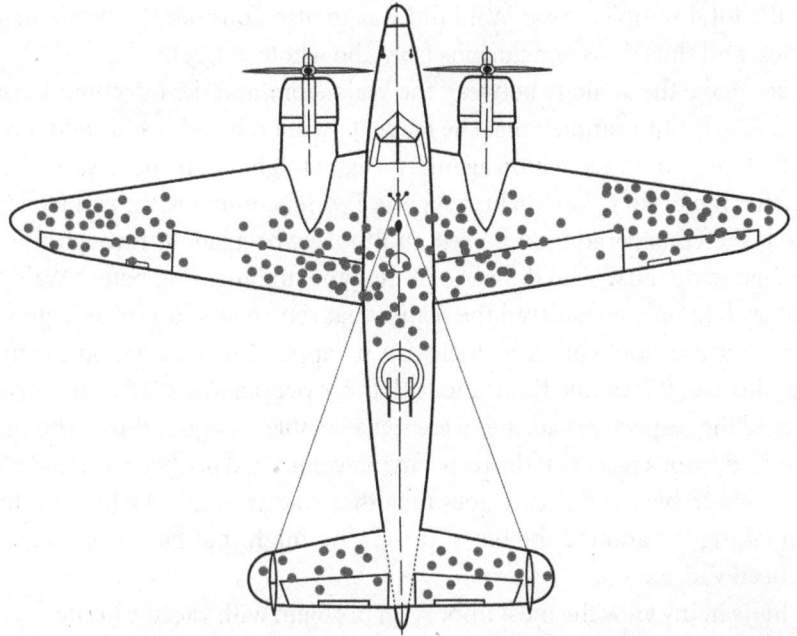

Figure 10.5 Illustration of hypothetical damage pattern on a World War II bomber. Wald suggested that the additional armor should be put wherever there were few or no bullet holes, because most of the planes that were hit there did not make it back (Creative Commons Attribution-Share Alike 4.0 International license; https://commons. wikimedia.org/wiki/File:Survivorship-bias.svg).

Wald, however, had a different opinion. He suggested that the additional armor should be put where there were very few bullet holes, most importantly the engines and the cockpit. Wald realized that there was no specific reason for the bullet holes to be concentrated in particular places; rather they should have been distributed evenly all over the plane. Therefore, he concluded that the engines and the cockpit had fewer bullet holes than other parts of the returning aircraft not because those parts had been less hit, but because those aircraft that had been hit in the engines and cockpit simply did not return. Indeed, and contrary to what had been initially thought, the planes returning with fuselage damage indicated that such hits could be tolerated. The attitude that led to the initial suggestion to reinforce with armor the fuselage but not the engines or the cockpit is often described as survivorship bias. It entails a selection bias that only considers the "survivors" of a particular process or event, while ignoring those who did not make it through. As a result, the conclusions are based on a partial sample (the "survivors"), and not the total sample. What Wald did was to also consider the nonreturning planes, and thus draw conclusions from the whole dataset.[20]

Let's make the analogy between the Wald story and the infectious disease. When Darla Shine argued that she made it through measles as a child, implying that therefore vaccination against measles might not be necessary, she fell into the trap of survivorship bias. As one Twitter commentator and Paul Offit noted, she considered only herself and those who made it through the disease, ignoring those who did not—in the same manner that before Wald, the military had only considered the planes that returned to their base, ignoring those that did not. The same logic can be applied to vaccination for infectious disease: We should think about all those people who died from a disease because the respective vaccines were not available, or about those who might have died from a disease if the respective vaccines had not been available. The survivorship bias in this case goes the other way around: The bias has to do with taking for granted the life of those who might not have lived had they not been vaccinated.

This is in my view the most important problem with vaccine hesitancy: We ask parents to insert a foreign element into the bodies of their babies, which might entail a risk (albeit a very small one), against a disease and a pathogen they do not see around them (with an admittedly higher risk, which is nevertheless theoretical). Even if they accept that the risks of vaccines are a lot less than the risk of the disease, they cannot be convinced to take the former risk when there is no immediate evidence of the latter. This is not unreasonable. Therefore, what they need to do, and the Wald story might help, is to realize that they do not see the higher risk of the disease because many people before them took the lower risk of being vaccinated. But the situation may change when more and more people refrain from being vaccinated against highly contagious pathogens such as the measles virus. And it might be too late if we end up letting the problems from the disease become more evident, because vaccines would not be that useful anymore.

Concluding remarks

Throughout the present book, I have presented what I consider some under-explored causes of science distrust. I do not underestimate how financial interests related to science as well as disinformation and misinformation about science may affect lay people's trust in it. Neither do I underestimate that scientific institutions and scientists themselves may not have done as much as necessary to gain and maintain lay people's trust. However, I argue that lay people need to be guided to better understand the questions that science can and cannot answer, as well as some of its features. If lay people do not have the necessary background to understand the features that I have discussed in the chapters of the present book, it will be hard to address their distrust. That said, I do not mean to suggest that lay people themselves are to blame for their distrust. Quite the contrary, if there is someone to blame, it is the educational systems that have not delivered the necessary understanding of these features of science. Instead, what most people likely remember from their science school years is being bombarded with content knowledge, which often made them wonder "why do I have to learn this stuff?" A key problem therefore has been the way that science has been taught in schools.

In a 2024 survey by the Edelman Trust Institute, people from 28 countries all over the world were asked to express their agreement or disagreement with the statement "Scientists do not know how to communicate with people like me." On average, about half of the participants agreed with this statement, ranging from about one in three in Japan to about two in three in China.[1] I take this as evidence of the lack of effective communication between scientists and the public. But in my view, this does not only happen because scientists do not communicate science as they should; this also happens because lay people have not been given the necessary tools to understand science. Once again, lay people are not to blame; the blame falls on educational systems and the teaching of science that is focused on science content knowledge and not on the nature of science. If lay people are not taught how science is done and what its potential and limitations are, it is no surprise that they feel that scientists' communication is ineffective.

Trusting Science. Kostas Kampourakis, Oxford University Press. © Oxford University Press (2025).
DOI: 10.1093/oso/9780197787106.003.0011

I like illustrating the problem with a map metaphor. Imagine that teachers ought to teach about geographical maps. What has been happening so far in science teaching in schools resembles the situation where teachers would ask their students to learn the contents of a map by heart (e.g., what the distances among cities are, what the altitude of particular areas is, etc.) and then test them on how well they have learned this information. This corresponds to asking students, for instance, to learn to represent the structure of the atom, to provide descriptions of cells and tissues, or to describe forces and interactions among objects. However, even if students managed to learn all that, they would likely forget most of the details after some time—ask yourself what you can recall from your school years. Instead of this, I suggest that what should be happening in schools should resemble the situation where teachers guide their students to understand what a map is (a model, or a representation of reality), and how to use it in their everyday life. This is most useful, because it is not rote knowledge learned by heart—it is a skill. Once students develop it, they are less likely to forget it. The information is nowadays available for people to find; what they need is to know how to use it. The problem then can be summarized by the question: "Learn the map, or how to use it?" If you replace the map with any scientific theory, concept, or explanation taught in schools, the suggested shift in science teaching in schools becomes evident. Instead of transmitting science content knowledge to students and asking them to learn it, we should help them learn how to use this knowledge in their everyday lives.

I am neither the first nor the only one to make this point. In a recent book, science educator and historian John Rudolph has cogently argued that we must at last drop the long-standing belief that a solid understanding of science depends on a solid knowledge base. All the available data from science education research show that this does not work. What we must do is entirely switch the focus of school science teaching, aiming at attaining two learning goals: The first is to help students understand the way scientists arrive at knowledge about the world, a shift we can describe as one from "what we know" to "how we came to know what we know." Rudolph argues, and I entirely agree, that by presenting the reasoning and the practices that have led scientists to the invention of concepts and theories, students will not only learn those concepts and theories but will also better understand them. Furthermore, this will help rebuild trust in science, as students will come to understand the real potential of science, as well as its limitations. The second goal is learning about the role of science in society. Instead of the lonely (White, male) genius stereotypical stories often found in textbooks, we must help students understand science as a social process that is in constant and mutual interaction with society. Students will thus be guided to understand that science is not something alien or external to society, but rather something coconstructed within it. Science and society influence each other in a variety of ways, and understanding why and how this happens is essential for understanding science.[2]

In the present book, I am making the same general argument with Rudolph, but I am also proposing more specific learning goals. Lay people are more likely to trust science if they better understand it and accept its limitations. My foray into the history of vaccination helped me realize particular reasons why lay people may distrust science. To change this, I suggest that lay people need to be able to understand:

1. How to distinguish between legitimate and illegitimate conspiracy theories when it comes to science.
2. How to distinguish between the scientific conclusions and the related ethical questions.
3. That uncertainty is inherent in science and that in our societies we will often have to choose between different risks.
4. How to deal with the conflicting claims of scientists, by distinguishing between real experts and mere "experts."
5. The internal processes of science and how to consider the findings of a field as a whole.
6. That science cannot "prove" anything, but only establish associations, or lack thereof, between variables.
7. How to distinguish between the scientific conclusions and the related political decisions.
8. How to make sense of the evidence for the claims of scientists, when it is not immediately observable.

The episodes discussed in the present book were presented in a roughly chronological order, and might seem distinct and not necessarily related. However, not only are they related, but they can also serve as distinct steps of a heuristic tool that can help both lay people to decide whether to trust scientists, and scientists to figure out what they need to do in order to gain public trust. These steps are summarized in the flowchart at the end of the present section. This heuristic tool can be used in order to evaluate all claims related to science and decide what to do about them.

The first question to ask is whether the claim under consideration (hereafter *the claim*) is related to a scientific question or not. As we saw, it is possible that the relevant questions are ethical (Chapter 4) or political (Chapter 9).

These are questions that can be informed by science but which science cannot itself answer. Rather, other nonscientific factors are crucial for the answers given and the relevant decisions made, even though these might draw at least partially on science. For instance, we know from science that to achieve herd immunity, most people should be vaccinated, but science tells us nothing about the moral obligation of fairness to do so by considering herd immunity as a public good. We also know what happens in fertilization, implantation, or embryo development, but science alone cannot tell us when personhood begins. At the same time, we may even decide to ignore science in making political decisions: People were sterilized in the name of eugenics in order to liberate human societies from allegedly defective mental traits; families have been compensated for vaccine injury despite the lack of scientific evidence that vaccines were indeed the cause of injury; and thimerosal was removed from vaccines despite the lack of scientific evidence that it was dangerous.

Once we decide that *the claim* is related to a scientific question, we may consider whether a conspiracy theory is related to that. As we saw, conspiracy theories derive from our natural tendency to look for patterns and agency around us, as well as from our feelings of fear and uncertainty. Therefore, rather than leave unaddressed the question of whether there is an underlying conspiracy, it is better to explicitly consider it. If we decide that there is not, we can move on. If we decide that there might be one, we should reject it if it is entirely implausible. If there is a chance that it might be plausible in principle, such as covering up that SARS-CoV-2 emerged from a lab leak, we should keep it in mind, but not get out of the logical path that brings us to the key question: Is *the claim* in agreement with the consensus view of the expert scientific community? This is where two diverging paths begin.

If the answer to this question is no, we do not need to outright reject *the claim*. There exist many cases in the history of science where scientists dissented from the mainstream view, offering new explanations for the natural phenomena observed (Darwin and Galileo would be the classic, though a bit stereotypical, examples). So, rather than rejecting *the claim*, we should ask whether the scientists supporting it are experts in this field. If they are not, then we should reject *the claim*. But if they are, we have good reasons to keep considering it. This would bring us to another key question, whether *the claim* has undergone peer review by experts in this field. If it has not, then we should reject it. But if it has, then we should keep it in mind. There are no serious reasons to question a claim made by experts in a field that has passed through peer review by their peers. Having done so would mean that the experts consider it legitimate, even if it is not the mainstream view, and so we should too.

If, in contrast, *the claim* agrees with the consensus view of the expert scientific community, then we have the best reasons to consider it. But there are three more elements that are necessary in order for us to trust scientists that *the claim* is true. These aspects are necessary not for the majority of expert scientists (this is when we have a consensus in the respective community) themselves to be sure about it; they are supposed to know what they are doing, and who are we to question them? Rather, this is for us to be able to understand it. The first aspect is for the expert scientific community to make clear what the available evidence for *the claim* is. The second aspect is for them to explain why the available evidence supports *the claim*. And the third aspect is for the expert scientific community to estimate and explain the risks related to *the claim*, if any (see Figure C1). Only when all these three pieces of information are in place, I argue, should we be able to fully trust scientists. Attention: I am not saying that we must not trust them otherwise; all I am suggesting is that we would have no qualms and no hesitation in trusting the expert scientific community if they have made clear what the available evidence for *the claim* is, why the available evidence supports it, and what the risks related to it, if any, are. If any of these pieces of information is missing, we should ask for it; perhaps even demand it.

These steps are represented in the flowchart below. I believe that it can be a very useful and easy-to-use heuristic tool for people to better understand science, and for scientists to fight distrust. Helping you, the reader, better understand what science is and how it is done is a crucial first step towards fighting science distrust. A second, long-term, but still key, step would be for this heuristic, or others like that, to be taught in schools. As recently argued, we should educate future citizens who will be "competent outsiders" able to evaluate the credibility of science-based arguments.[3] I hope that the present book will make a modest contribution to this end, and that it can help lay people better understand why and when to trust science.

What about the future? We need to reconsider science communication, and encourage the active participation of scientists in it, in order to enhance the public understanding of science. Instead of having journalists supposedly representing different views "equally," the consensus positions of the scientific community for various topics should be communicated clearly and effectively via independent and reliable venues that directly represent the scientific community itself. This requires the reorganization of academia itself, so that resources are devoted to science communication and incentives are given to scientists taking an active part in it. This has to become professionalized in the same way that science education is. Science departments and institutions must have experts in science communication and public engagement who will be aware of the public views and attitudes and who will be able to present new findings and conclusions in a way that enhances laypersons' understanding.

We need a society that relies on science by trusting it, while also understanding and accepting its limitations. In science we trust, but not blindly. Science relies on understanding and not dogmatic belief; therefore, only if we understand how it is done can we decide to trust it. I hope that the heuristic tool on the next page will help you decide when to trust scientific claims.

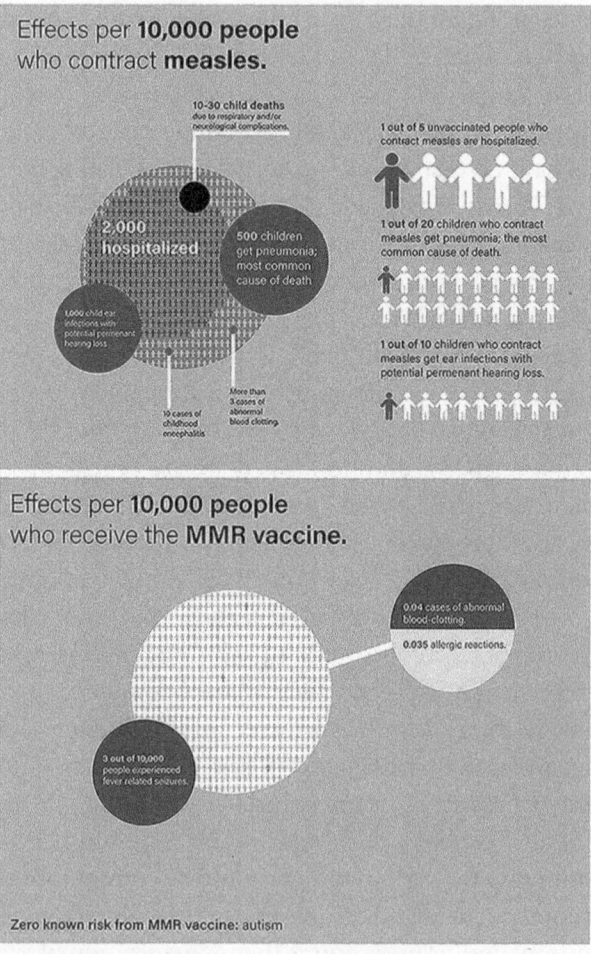

Figure C1 Comparing the risk of 10,000 children acquiring measles versus receiving the measles-mumps-rubella (MMR) vaccine. Reprinted from Hotez, P. (2025). It won't end with COVID: Countering the next phase of American antivaccine activism 2025–29. PLOS *Glob Public Health*, 5(1), e0004020 (CC BY 4.0).

When to trust scientific claims

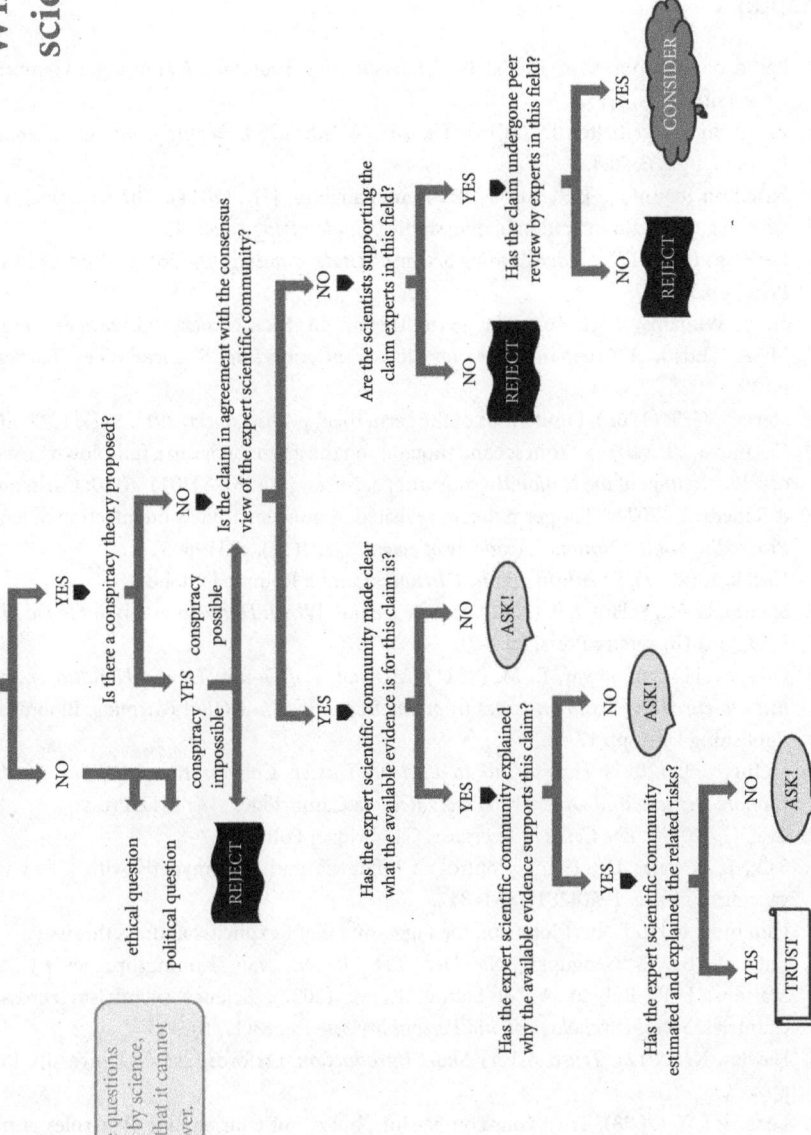

References

Chapter 1

1. Based on Kampourakis, K. (2020). *Understanding Evolution*. Cambridge: Cambridge University Press, p.152.
2. Based on Spiegelhalter, D. (2019). *The Art of Statistics: Learning from Data*. London: Penguin, pp.283–284.
3. Based on Ioannidis, J. P., Tarone, R., & McLaughlin, J. K. (2011). The false-positive to false-negative ratio in epidemiologic studies. *Epidemiology*, 450–456.
4. De Regt, H. (2017). *Understanding Scientific Understanding*. Oxford: Oxford University Press, pp.37–40.
5. Steere-Williams, J. (2016). The germ theory. In Montgomery, G. M., & Largent, M. A. (Eds.), *A Companion to the History of American Science*. Wiley Blackwell, pp.397–407.
6. Stewart, G. T. (1968). Limitations of the germ theory. *The Lancet*, *291*(7551), 1077–1081.
7. Casanova, J. L. (2023). From second thoughts on the germ theory to a full-blown host theory. *Proceedings of the National Academy of Sciences*, *120*(26), e2301186120; Carlsson, F., & Råberg, L. (2024). The germ theory revisited: A noncentric view on infection outcome. *Proceedings of the National Academy of Sciences*, *121*(17), e2319605121.
8. Haack, S. (2017). *Scientism and Its Discontents*. n.p.: Rounded Globe.
9. Sinatra, G. M., & Hofer, B. (2021). *Science Denial: Why It Happens and What to Do About It*. Oxford University Press, pp.9–10.
10. Oreskes, N., & Conway, E. M. (2011). *Merchants of Doubt: How a Handful of Scientists Obscured the Truth on Issues from Tobacco Smoke to Global Warming*. Bloomsbury Publishing USA, pp.12–15.
11. McIntyre, L. (2021). *How to Talk to a Science Denier: Conversations with Flat Earthers, Climate Deniers, and Others Who Defy Reason*. Cambridge, MA: MIT Press.
12. Eyal, G. (2019). *The Crisis of Expertise*. Cambridge: Polity, pp.7–8.
13. Salk, J., & Salk, D. (1977). Control of influenza and poliomyelitis with killed virus vaccines. *Science*, *195*(4281), 834–847.
14. I am indebted to Robert Johnston for suggesting that I explicitly address this issue.
15. Rutjens, B. T., Sengupta, N., Der Lee, R. V., van Koningsbruggen, G. M., Martens, J. P., Rabelo, A., & Sutton, R. M. (2022). Science skepticism across 24 countries. *Social Psychological and Personality Science*, *13*(1), 102–117.
16. Hawley, K. (2012). *Trust: A Very Short Introduction*. Oxford: Oxford University Press, pp.4–7.
17. Lenard, P. T. (2008). Trust your compatriots, but count your change: The roles of trust, mistrust and distrust in democracy. *Political Studies*, *56*(2), 312–332.
18. Griffith, D. M., Bergner, E. M., Fair, A. S., & Wilkins, C. H. (2021). Using mistrust, distrust, and low trust precisely in medical care and medical research advances health equity. *American Journal of Preventive Medicine*, *60*(3), 442–445.

19. The fact that science denialists make claims that are factually wrong makes a priori not an entirely accurate descriptor, but I use it here because it helps make the distinction between theoretical assumptions and experience.

20. Collins, H. (2014). *Are We All Scientific Experts Now?* Cambridge: Polity, pp.49–74.

21. Nichols, T. (2017). *The Death of Expertise: The Campaign Against Established Knowledge and Why It Matters.* Oxford: Oxford University Press, pp.28–39.

22. Oreskes, N., & Conway, E. M. (2011). Merchants of Doubt: How A Handful of Scientists Obscured the Truth on Issues from Tobacco Smoke to Global Warming. Bloomsbury Publishing USA, pp.12–15.

23. Romm, J. (2022). *Climate Change: What Everyone Needs to Know*, 3rd edition. New York: Oxford University Press.

24. Hardwig, J. (1991). The role of trust in knowledge. *The Journal of Philosophy*, *88*(12), 693–708.

25. De Ridder, J. (2022). How to trust a scientist. *Studies in History and Philosophy of Science*, *93*, 11–20.

26. Ipsos Global Trustworthiness Index. 2022. https://www.ipsos.com/en-us/news-polls/global-trustworthiness-index-2022.

27. Wellcome Monitor. (2020). Wellcome Global Monitor. How Covid-19 Affected People's Lives and Their Views About Science. https://wellcome.org/reports/wellcome-global-monitor-covid-19/2020.

28. 2024 Edelman Trust Barometer Global Report. https://www.edelman.com/trust/2024/trust-barometer.

29. Eyal, G. (2019). *The Crisis of Expertise.* Cambridge: Polity, Chapter 3; see also Kampourakis, K. (2020). *Understanding Evolution.* Cambridge: Cambridge University Press for a similar analysis of polls on evolution acceptance.

30. Gligorić, V., van Kleef, G. A., & Rutjens, B. T. (2024). How social evaluations shape trust in 45 types of scientists. *PLoS ONE*, *19*(4), e0299621.

31. Wellcome Global Monitor 2020. https://wellcome.org/reports/wellcome-global-monitor-covid-19/2020.

32. Newman-Toker, D. E., Nassery, N., Schaffer, A. C., Yu-Moe, C. W., Clemens, G. D., Wang, Z., ... & Siegal, D. (2024). Burden of serious harms from diagnostic error in the USA. *BMJ Quality & Safety*, *33*(2), 109–120.

33. Auerbach, A. D., Lee, T. M., Hubbard, C. C., Ranji, S. R., Raffel, K., Valdes, G., ... & UPSIDE Research Group. (2024). Diagnostic errors in hospitalized adults who died or were transferred to intensive care. *JAMA Internal Medicine.* doi:10.1001/jamainternmed.2023.7347

Chapter 2

1. MacKenzie, D. (2020). *Covid-19—The Pandemic That Never Should Have Happened, and How to Stop the Next One.* London: The Bridge Street Press; Horton, R. (2020). *The Covid—19 Catastrophe: What's Gone Wrong and How to Stop It Happening Again.* Cambridge: Polity.

2. Rabadan, R. (2021). *Understanding Coronavirus.* Cambridge: Cambridge University Press.

3. Elgar, F. J., Lahti, H., Lopes Ferreira, J., Melkumova, M., & Bilz, L. *Navigating Uncharted Territory: School Closures and Adolescent Experiences During the COVID-19 Pandemic in the WHO European Region. Impact of the COVD-19 Pandemic on Young People's Health and Well-Being from the Findings of the HBSC Survey Round 2021/22.* Copenhagen: WHO Regional Office for Europe; 2023. https://apps.who.int/iris/handle/10665/369723.

4. https://www.aei.org/covid-19-education-response-longitudinal-survey-c-erls/ (accessed December 21, 2023).

5. https://www.cityofmillvalley.org/902/Stop-5-Locust-Miller (accessed November 30, 2023).

6. Spinney, L. (2017). *Pale Rider: The Spanish Flu of 1918 and How It Changed the World.* London: Vintage, pp.3–5, 61–63.

7. Markel, H., Lipman, H. B., Navarro, J. A., Sloan, A., Michalsen, J. R., Stern, A. M., & Cetron, M. S. (2007). Nonpharmaceutical interventions implemented by US cities during the 1918-1919 influenza pandemic. *Journal of the American Medical Association, 298*(6), 644–654; Correia, S., Luck, S., & Verner, E. (2022). Pandemics depress the economy, public health interventions do not: Evidence from the 1918 flu. *The Journal of Economic History, 82*(4), 917–957.

8. Correia, S., Luck, S., & Verner, E. (2022). Pandemics depress the economy, public health interventions do not: Evidence from the 1918 flu. *The Journal of Economic History, 82*(4), 917–957.

9. Jenner, E. (1798). *An Inquiry into the Causes and Effects of the Variolae Vaccinae: A Disease Discovered in Some of the Western Counties of England, Particularly Gloucestershire, and Known by the Name of the Cow Pox.* London: Sampson Low/ Murray and Highly, p.45.

10. Dunning, R. (1880). *Some Observations on Vaccination or the Inoculated Cow-Pox.* London: March and Teape.

11. Pasteur, M. (1881). An address on vaccination in relation to chicken cholera and splenic fever. *British Medical Journal, 2*(1076), 283–284, p.284.

12. Dai, L., & Gao, G. F. (2021). Viral targets for vaccines against COVID-19. *Nature Reviews Immunology, 21*(2), 73–82.

13. Based on Dowdle, W. R. (1998). The principles of disease elimination and eradication. *Bulletin of the World Health Organization, 76*(Suppl 2), 22. https://www.nlm.nih.gov/oet/ed/stats/01-200.html#:~:text=Incidence%20and%20prevalence%20represent%20measures,are%20measures%20of%20disease%20burden (accessed December 2, 2023).

14. Wu, N., Joyal-Desmarais, K., Ribeiro, P. A., Vieira, A. M., Stojanovic, J., Sanuade, C., ... & Bacon, S. L. (2023). Long-term effectiveness of COVID-19 vaccines against infections, hospitalisations, and mortality in adults: Findings from a rapid living systematic evidence synthesis and meta-analysis up to December, 2022. *The Lancet Respiratory Medicine, 11*(5), 439–452.

15. Goldenberg, M. J. (2021). *Vaccine Hesitancy: Public Trust, Expertise, and the War on Science.* Pittsburgh: University of Pittsburgh Press, pp.3–4.

16. BMJ 2021;372:n597 http://dx.doi.org/10.1136/bmj.n597

17. https://wellcome.org/reports/wellcome-global-monitor/2018/chapter-5-attitudes-vaccines# (accessed January 16, 2024).

18. Lazarus, J. V., Wyka, K., White, T. M., Picchio, C. A., Rabin, K., Ratzan, S. C., ... & El-Mohandes, A. (2022). Revisiting COVID-19 vaccine hesitancy around the world using data from 23 countries in 2021. *Nature Communications, 13*(1), 3801.

19. Goldenberg, M. J. (2021). *Vaccine Hesitancy: Public Trust, Expertise, and the War on Science.* Pittsburgh: University of Pittsburgh Press, pp.129–135, 138–140.

20. Goldacre, B. (2013). *Bad Pharma: How Medicine in Broken and How to Fix It.* London: Fourth Estate.

21. Almashat, S., Wolfe, S. M., & Carome, M. (2016). Twenty-five years of pharmaceutical industry criminal and civil penalties: 1991 through 2015. *Public Citizen.* https://www.citizen.org/article/twenty-five-years-of-pharmaceutical-industry-criminal-and-civil-penalties-1991-through-2015/ (accessed January 15, 2024).

22. Lyman, S. (2019). Pharma's tarnished reputation helps fuel the anti-vaccine movement. *Stat News.* https://www.statnews.com/2019/02/26/anti-vaccine-movement-pharma-tarnished-reputation/. (accessed January 15, 2024).

23. Grignolio, A. (2018). *Vaccines: Are They Worth a Shot?* Cham: Springer, pp.46–52.

24. https://www.cnbc.com/2023/01/31/the-covid-pandemic-drives-pfizers-2022-revenue-to-a-record-100-billion.html (accessed January 22, 2024).

25. McIntyre, L. (2023). *On Disinformation: How to Fight for Truth and Protect Democracy.* Cambridge, MA: MIT Press.

26. Hornsey, M. J., Finlayson, M., Chatwood, G., & Begeny, C. T. (2020). Donald Trump and vaccination: The effect of political identity, conspiracist ideation and presidential tweets on vaccine hesitancy. *Journal of Experimental Social Psychology, 88,* 103947.

27. McIntyre, L. (2023). *On Disinformation: How to Fight for Truth and Protect Democracy.* Cambridge, MA: MIT Press; van der Linden, S. (2023). *Foolproof: Why Misinformation Infects Our Minds and How to Build Immunity.* London: Fourth Estate.

28. Goldenberg, M. J. (2021). *Vaccine Hesitancy: Public Trust, Expertise, and the War on Science.* Pittsburgh: University of Pittsburgh Press, pp.31–32.

29. Kampourakis, K. (2016). The "general aspects" conceptualization as a pragmatic and effective means to introducing students to nature of science. *Journal of Research in Science Teaching, 53*(5), 667–682.

Chapter 3

1. Reverby, S. M. (Ed.). (2000). Tuskegee's *Truths: Rethinking the Tuskegee Syphilis Study.* Chapel Hill, NC: University of North Carolina Press, p.136.

2. The story presented in this chapter is based on Jones, J. H. (1993/1981). *Bad Blood: The Tuskegee Syphilis Experiment.* New York: The Free Press, and Reverby, S. M. (2009). *Examining Tuskegee: The Infamous Syphilis Study and Its Legacy.* Chapel Hill, NC: University of North Carolina Press. Unless otherwise mentioned, the original documents from which the quotations come are reprinted in Reverby, S. M. (Ed.). (2000). Tuskegee's *Truths: Rethinking the Tuskegee Syphilis Study.* Chapel Hill, NC: University of North Carolina Press.

3. https://www.nytimes.com/1972/07/26/archives/syphilis-victims-in-us-study-went-untreated-for-40-years-syphilis.html (accessed January 13, 2024).

4. Zimmerman, E. L. (1921). A comparative study of syphilis in Whites and in Negroes. *Archives of Dermatology and Syphilology, 4,* 75–88, p.76.

5. Zimmerman, E. L. (1921). A comparative study of syphilis in Whites and in Negroes. *Archives of Dermatology and Syphilology*, 4, 75–88, p.79.

6. Jones, J. H. (1993/1981). *Bad Blood: The Tuskegee Syphilis Experiment*. New York: The Free Press, p.94.

7. Lombardo, P. A., & Dorr, G. M. (2006). Eugenics, medical education, and the Public Health Service: Another perspective on the Tuskegee syphilis experiment. *Bulletin of the History of Medicine*, 291–316.

8. Jones, J. H. (1993/1981). *Bad Blood: The Tuskegee Syphilis Experiment*. New York: The Free Press, p.163, emphasis in the original.

9. Reverby, S. M. (2022). "An Opportunity of This Kind": The Milbank Memorial Fund and the U.S. Public Health Service Study of Untreated Syphilis in Tuskegee. A Report to the Milbank Memorial Fund. https://www.milbank.org/wp-content/uploads/2022/04/Reverby_report_4.24.pdf.

10. Heller Jr., J. R., & Bruyere, P. T. (1946). Untreated syphilis in the male Negro. II. Mortality during 12 years of observation. *Journal of Venereal Disease Information*, 27(2), 34–38.

11. Jones, J. H., & Reverby, S. M. (2022). 50 Years after the Tuskegee revelations: Why does the mistrust linger? *American Journal of Public Health*, 112(11), 1538–1540.

12. https://clintonwhitehouse4.archives.gov/textonly/New/Remarks/Fri/19970516-898.html#:~:text=To%20Macon%20County%2C%20to%20Tuskegee,be%20allowed%20to%20happen%20again.

13. Jones, J. H., & Reverby, S. M. (2022). 50 Years after the Tuskegee revelations: Why does the mistrust linger? *American Journal of Public Health*, 112(11), 1538–1540, p.1359.

14. https://www.cdc.gov/tuskegee/after.htm (accessed February 2, 2024)

15. Washington, H. A. (2008/2006). *Medical Apartheid: The Dark History of Medical Experimentation on Black Americans from Colonial Times to the Present*. New York: Anchor Books, p.386.

16. Bajaj, S. S., & Stanford, F. C. (2021). Beyond Tuskegee—vaccine distrust and everyday racism. *New England Journal of Medicine*, 384(5), e12.

17. Oliver, J. E., & Wood, T. (2014). Medical conspiracy theories and health behaviors in the United States. *JAMA Internal Medicine*, 174(5), 817–818.

18. Cassam, Q. (2019). *Conspiracy Theories*. Cambridge: Polity, pp.3, 6, 11.

19. Nattrass, N. (2012). *The AIDS Conspiracy: Science Fights Back*. New York: Columbia University Press.

20. https://twitter.com/Bryce_Nickels/status/1753090035950432657 (accessed February 2, 2024).

21. https://twitter.com/DrJBhattacharya/status/1753269342907875418 (accessed February 2, 2024).

22. See, for instance, Andersen, K. G., Rambaut, A., Lipkin, W. I., Holmes, E. C., & Garry, R. F. (2020). The proximal origin of SARSCoV2. *Nature Medicine*, 26(4), 450–452; Lu, R., Zhao, X., Li, J., Niu, P., Yang, B., Wu, H., Wang, W., Song, H., Huang, B., Zhu, N., & Bi, Y. (2020). Genomic characterisation and epidemiology of 2019 novel coronavirus: implications for virus origins and receptor binding. *The Lancet*, 395(10224), 565–574.

23. Worobey, M. (2021). Dissecting the early COVID-19 cases in Wuhan. *Science*, 374(6572), 1202–1204.

24. Lewandowsky, S., Jacobs, P. H., & Neil, S. (2023). Leak or leap? Evidence and cognition surrounding the origins of the SARS-CoV-2 virus. In Butter, M., & Knight, P. (Eds.), *Covid Conspiracy Theories in Global Perspective*, 26–39. London: Routledge.

25. Ackerman, G., Behlendorf, B., Baum, S., Peterson, H., Wetzel, A., & Halstead, J. (2024). *The Origin and Implications of the COVID-19 Pandemic: An Expert Survey*. Global Catastrophic Risk Institute Technical Report 24-1.

26. Uscinski, J. E., Enders, A. M., Klofstad, C. A., Seelig, M. I., Funchion, J. R., Everett, C., Wuchty, S., Premaratne, K., & Murthi, M. N. (2020). Why do people believe COVID-19 conspiracy theories? *Harvard Kennedy School (HKS) Misinformation Review*.

27. Regazzi, L., Lontano, A., Cadeddu, C., Di Padova, P., & Rosano, A. (2023). Conspiracy beliefs, COVID-19 vaccine uptake and adherence to public health interventions during the pandemic in Europe. *European Journal of Public Health*, *33*(4), 717–724.

28. Salman, M., Mallhi, T., Tanveer, N., Shehzadi, N., Khan, H., Ul Mustafa, Z., ... & Khan, Y. H. (2022). Evaluation of conspiracy beliefs, vaccine hesitancy, and willingness to pay towards COVID-19 vaccines in six countries from Asian and African regions: a large multinational analysis. *Vaccines*, *10*(11), 1866.

29. Caycho-Rodríguez, T., Ventura-León, J., Valencia, P. D., Vilca, L. W., Carbajal-León, C., Reyes-Bossio, M., White, M., Rojas-Jara, C., Polanco-Carrasco, R., Gallegos, M., Cervigni, M., Martino, P., Palacios, D. A., Moreta-Herrera, R., Samaniego-Pinho, A., Lobos Rivera, M. E., Buschiazzo Figares, A., Puerta-Cortés, D. X., Corrales-Reyes, I. E., Calderón, R., Pinto Tapia, B., Arias Gallegos, W. L., & Petzold, O. (2022). What is the support for conspiracy beliefs about COVID-19 vaccines in Latin America? A prospective exploratory study in 13 countries. *Frontiers in Psychology*, *13*, 855713.

30. Van Prooijen, J.-H. (2018). *The Psychology of Conspiracy Theories*. New York: Routlegde, pp.5–6.

31. Longerich, P. (2021). *Wannsee: The Road to the Final Solution*. Oxford: Oxford University Press.

32. Barkun, M. (2013). *A Culture of Conspiracy: Apocalyptic Visions in Contemporary America*. Berkeley: University of California Press, pp.3–4.

33. Barkun, M. (2013). *A Culture of Conspiracy: Apocalyptic Visions in Contemporary America*. Berkeley: University of California Press, p.6.

34. See Gagné, M. J. (2022). *Thiking Critically about the Kennedy Assasination: Debunking the Myths and Conspiracy Theories*. New York: Routledge.

35. Evans, R. J. (2021). *The Hitler Conspiracies: The Third Reich and the Paranoid Imagination*. Penguin.

36. Cassam, Q. (2019). *Conspiracy Theories*. Cambridge: Polity, pp.16–28, 97–98.

37. Van Prooijen, J.-H. (2018). *The Psychology of Conspiracy Theories*. New York: Routlegde, p.89.

38. Gagné, M. J. (2022). *Thiking Critically about the Kennedy Assasination: Debunking the Myths and Conspiracy Theories*. New York: Routledge, p. xviii.

Chapter 4

1. https://www.nationalarchives.gov.uk/education/resources/victorian-health-reform/ source-3-letter-protesting-vaccination-bill/ (accessed February 9, 2024).

2. The story presented in this chapter is based on Moulin, A. M. (dir.), *L'Aventure de la Vaccination*, Paris, Fayard; Williams, G. (2010). *Angel of Death: The Story of Smallpox*. London: Palgrave Macmillan; Durbach, N. (2004). *Bodily Matters: The Anti-Vaccination Movement in England, 1853–1907*. Durham, NC: Duke University Press; Williamson, S. (2007).

The Vaccination Controversy: The Rise, Reign and Fall of Compulsory Vaccination for Smallpox. Liverpool: Liverpool University Press.

3. Jenner, E. (1798). *An Inquiry into the Causes and Effects of the Variolae Vaccinae: A Disease Discovered in Some of the Western Counties of England, Particularly Gloucestershire, and Known by the Name of the Cow Pox.* London: Sampson Low/ Murray and Highly, p.45.

4. Moseley, B. (1800). *A Treatise on Sugar: With Miscellaneous Medical Observations.* London: Printed by John Nichols, p.183.

5. https://www.britishmuseum.org/collection/object/P_1851-0901-1091; https://www.themorgan.org/blog/cow-pock-or-wonderful-effects-new-inoculation (accessed February 9, 2024).

6. Guy, W. A. (1882). Two hundred and fifty years of small pox in London. *Journal of the Statistical Society of London, 45*(3), 399–443.

7. Small pox and vaccination. Copy of letter from Dr. Edward Seaton to Viscount Palmerston, with enclosed copy of a report on the state of small pox and vaccination in England and Wales and other countries, and on compulsory vaccination, p.21. https://wellcomecollection.org/works/gcmc9hwj (accessed February 9, 2024)

8. (1853) An act (16 & 17 Vict. cap. 100.) further to extend and make compulsory the practice of vaccination: passed August 20, 1853. London: Printed by George E. Eyre and William Spottiswoode. https://archive.org/details/b22279714.

9. Gibbs, J. (1854). *Our Medical Liberties, or the Personal Rights of the Subject, as Infringed by Recent and Proposed Legislation: Compromising Observations on the Compulsory Vaccination Act, the Medical Registration and Reform Bills, and the Maine Law.* London: Sotheran, Son, and Draper, p.7. https://wellcomecollection.org/works/rj45ue7g.

10. (1867). An Act to consolidate and amend the laws relating to vaccination. London: Printed by Eyre and Spottiswoode. https://archive.org/details/b3047579x.

11. Anti-Compulsory Vaccination League. (1867). *The British Medical Journal, 1*(323), 273.

12. Guy, W. A. (1882). Two hundred and fifty years of small pox in London. *Journal of the Statistical Society of London, 45*(3), 399–443.

13. Navin, M. (2016). *Values and Vaccine Refusal: Hard Questions in Ethics, Epistemology, and Health Care.* New York: Routledge, pp.99–103.

14. Field, R. I., & Caplan, A. L. (2008). A proposed ethical framework for vaccine mandates: Competing values and the case of HPV. *Kennedy Institute of Ethics Journal, 18*(2), 111–124.

15. Driver, J., The history of utilitarianism. In Zalta, E. N., & Nodelman, U. (Eds.), *The Stanford Encyclopedia of Philosophy* (Winter 2022 Edition). URL = <https://plato.stanford.edu/archives/win2022/entries/utilitarianism-history/>.

16. Field, R. I., & Caplan, A. L. (2008). A proposed ethical framework for vaccine mandates: competing values and the case of HPV. *Kennedy Institute of Ethics Journal, 18*(2), 111–124.

17. Navin, M. (2016). *Values and Vaccine Refusal: Hard Questions in Ethics, Epistemology, and Health Care.* New York: Routledge, pp.136–140.

18. Navin, M. (2016). *Values and Vaccine Refusal: Hard Questions in Ethics, Epistemology, and Health Care.* New York: Routledge, pp.140–143.

19. Giubilini, A. (2019). *The Ethics of Vaccination.* Cham: Springer/Palgrave Pivot, pp.95–103.

20. Davies, J. A. (2014). *Life Unfolding: How the Human Body Creates Itself.* Oxford: Oxford University Press, pp.38–40.

21. Jacobs, S. A. (2021). The scientific consensus on when a human's life begins. *Issues in Law & Medicine, 36,* 221–233, p.230.

22. Gilbert, S. F. (2023). Pseudo-embryology and personhood: How embryological pseudo-science helps structure the American abortion debate. *Natural Sciences, 3*(1), e20220041; see also Gilbert, S. F. (2008). When "personhood" begins in the embryo: Avoiding a syllabus of errors. *Birth Defects Research Part C: Embryo Today: Reviews, 84*(2), 164–173.

23. Gilbert, S. F. (2023). Pseudo-embryology and personhood: How embryological pseudo-science helps structure the American abortion debate. *Natural Sciences, 3*(1), e20220041, p.2.

24. Sarkar, S. (2021). Defining when human life begins is not a question science can answer—it's a question of politics and ethical values. *The Conversation.* https://theconversation.com/defining-when-human-life-begins-is-not-a-question-science-can-answer-its-a-question-of-politics-and-ethical-values-165514 (accessed June 30, 2024).

25. Jacobs, S. A. (2021). The scientific consensus on when a human's life begins. *Issues in Law & Medicine, 36,* 221–233, p.228.

26. Oreskes, N. (2021). *Why Trust Science?* Princeton: Princeton University Press, p.64.

27. See Part II of Kampourakis, K. (2018). *Turning Points: How Critical Events Have Driven Human Evolution, Life and Development.* Amherst, NY: Prometheus Books.

28. Rai, R., & Regan, L. (2006). Recurrent miscarriage. *The Lancet, 368*(9535), 601–611.

29. Fretts, R. C., Schmittdiel, J., McLean, F. H., Usher, R. H., & Goldman, M. B. (1995). Increased maternal age and the risk of fetal death. *New England Journal of Medicine, 333*(15), 953–957.

30. Andersen, A. M. N., Wohlfahrt, J., Christens, P., Olsen, J., & Melbye, M. (2000). Maternal age and fetal loss: Population based register linkage study. *BMJ, 320*(7251), 1708–1712.

31. van den Berg, M. M., van Maarle, M. C., van Wely, M., & Goddijn, M. (2012). Genetics of early miscarriage. *Biochimica et Biophysica Acta (BBA)-Molecular Basis of Disease, 1822*(12), 1951–1959.

32. Fragouli, E., Alfarawati, S., Spath, K., Jaroudi, S., Sarasa, J., Enciso, M., & Wells, D. (2013). The origin and impact of embryonic aneuploidy. *Human Genetics, 132,* 1001–1013.

33. Shattock, A. J., Johnson, H. C., Sim, S. Y., Carter, A., Lambach, P., Hutubessy, R. C., ... & Bar-Zeev, N. (2024). Contribution of vaccination to improved survival and health: Modelling 50 years of the Expanded Programme on Immunization. *The Lancet, 403*(10441), 2307–2316.

34. The moral basis for taxation is a complicated issue, and its detailed discussion falls outside the scope of the present chapter. See Frecknall-Hughes, J. (2019). The moral basis for taxation. In van Brederode, R. F. (Ed.), *Ethics and Taxation.* Springer, Singapore, pp.23–45.

35. Reich, J., & Gross, A. S. (2022, April). Vaccine hesitancy and individualism: An interview with Jennifer Reich. *New American Studies Journal: A Forum, 72,* 10.

36. Reich, J., & Gross, A. S. (2022, April). Vaccine hesitancy and individualism: An interview with Jennifer Reich. *New American Studies Journal: A Forum, 72,* 9.

37. Betsch, C., Böhm, R., Korn, L., & Holtmann, C. (2017). On the benefits of explaining herd immunity in vaccine advocacy. *Nature Human Behaviour, 1*(3), 0056.

Chapter 5

1. Oshinsky, D. M. (2005). *Polio: An American Story*. Oxford: Oxford University Press, p.195. Whenever there are no specific references, the details come from these books.
2. The story presented in this chapter is based on Smith, J. S. (1990). *Patenting the Sun: Polio and the Salk Vaccine*. William Morrow; Oshinsky, D. M. (2005). *Polio: An American Story*. Oxford: Oxford University Press; Offit, P. A. (2005). *The Cutter Incident: How America's First Polio Vaccine Led to the Growing Vaccine Crisis*. New Haven: Yale University Press; and Williams, G. (2013). *Paralysed with Fear: The Story of Polio*. New York: Palgrave.
3. https://content.time.com/time/subscriber/article/0,33009,860135,00.html   (accessed November 22, 2023).
4. Milzer, A., Levinson, S. O., Shaughnessy, H. J., Janota, M., Vanderboom, K., & Oppen-heimer, F. (1954). Immunogenicity studies in human subjects of trivalent tissue culture poliomyelitis vaccine inactivated by ultraviolet irradiation. *American Journal of Public Health*, 44(1), 26–33.
5. https://content.time.com/time/subscriber/article/0,33009,819686,00.html   (accessed November 22, 2023).
6. Oshinsky, D. M. (2005). *Polio: An American Story*. Oxford: Oxford University Press, p.197.
7. https://www.cdc.gov/polio/what-is-polio/index.htm (accessed November 24, 2023)
8. For an overview, see: Pollard, A. J., & Bijker, E. M. (2021). A guide to vaccinology: from basic principles to new developments. *Nature Reviews Immunology*, 21(2), 83–100.
9. Kolmer, J. A. (1936). Vaccination against acute anterior poliomyelitis. *American Journal of Public Health*, 26(2), 126–135.
10. Rivers, T. M. (1936). Immunity in virus diseases with particular reference to poliomyelitis. *American Journal of Public Health*, 26(2), 136–142.
11. Brodie, M., & Park, W. H. (1936). Active immunization against poliomyelitis. *American Journal of Public Health*, 26(2), 119–125.
12. Rivers, T. M. (1936). Immunity in virus diseases with particular reference to poliomyelitis. *American Journal of Public Health*, 26(2), 136–142.
13. Flexner, S. (1935). Concerning active immunization in poliomyelitis. *Science*, 82(2131), 420–421.
14. Enders, J. (1954). Some recent advances in the study of poliomyelitis. *Medicine*, 33, 87–95.
15. Quoted in Oshinsky, D. M. (2005). *Polio: An American Story*. Oxford: Oxford University Press, p.179.
16. Salk, J. E., Bennett, B. L., Lewis, L. J., Ward, E. N., & Youngner, J. S. (1953). Studies in human subjects on active immunization against poliomyelitis: 1. A preliminary report of experiments in progress. *Journal of the American Medical Association*, 151(13), 1081–1098.
17. Based on Grimes, D. A., & Schulz, K. F. (2002). An overview of clinical research: The lay of the land. *The Lancet*, 359(9300), 57–61.
18. Meldrum, M. (1998). "A calculated risk": The Salk polio vaccine field trials of 1954. *BMJ*, 317(7167), 1233–1236.
19. Quoted in Dawson, L. (2004). The Salk Polio Vaccine Trial of 1954: Risks, randomization and public involvement in research. *Clinical Trials*, 1(1), 122–130.
20. Benison, S. (1967). *Tom Rivers: Reflections on a Life in Medicine and Science (An Oral History Memoir)*. Cambridge, MA: MIT Press, p.497.

21. https://polioeradication.org/polio-today/polio-now/ (accessed November 24, 2023).

22. https://www.who.int/docs/default-source/documents/gpei-cvdpv-factsheet-march-2017.
    pdf?sfvrsn=1ceef4af2#:~:text=On%20very%20rare%20occasions%2C%20in,usually%
    20at%20the%20first%20dose (accessed November 24, 2023).

23. Kampourakis, K. & McCain, K. (2019). *Uncertainty: How It Makes Science Advance*. New York: Oxford University Press.

24. Gigerenzer, G. (2002). *Calculated Risks: How to Know When Numbers Deceive You*. New York: Simon and Schuster, p.26.

25. Gigerenzer, G. (2014). *Risk Savvy: How to Make Good Decisions*. New York: Viking, p.23.

26. https://www.who.int/news-room/questions-and-answers/item/cancer-carcinogenicity-of-the-consumption-of-red-meat-and-processed-meat (accessed November 24, 2023).

27. International Agency for Research on Cancer. (2018). Red Meat and Processed Meat Volume 114. IARC Monographs on the Evaluation of Carcinogenic Risks to Humans. Lyon, France. (p.493).

28. Ellis, P. D. (2010). *The Essential Guide to Effect Sizes: Statistical Power, Meta-Analysis, and the Interpretation of Research Results*. Cambridge: Cambridge University Press, Chapter 5.

29. Spiegelhalter, D. (2019). *The Art of Statistics: Learning from Data*. London: Penguin, pp.31–36.

30. https://www.who.int/docs/default-source/documents/gpei-cvdpv-factsheet-march-2017.pdf?sfvrsn=1ceef4af2#:~:;text=On%20very%20rare%20occasions%2C%20in, usually%20at%20the%20first%20dose (accessed November 24, 2023).

31. Cooper, L. V., Bandyopadhyay, A. S., Gumede, N., Mach, O., Mkanda, P., Ndoutabé, M., ... & Blake, I. M. (2022). Risk factors for the spread of vaccine-derived type 2 polioviruses after global withdrawal of trivalent oral poliovirus vaccine and the effects of outbreak responses with monovalent vaccine: a retrospective analysis of surveillance data for 51 countries in Africa. *The Lancet Infectious Diseases, 22*(2), 284–294.

32. https://polioeradication.org/this-week/variant-poliovirus-cvdpv/#:~:;text=Over%20the%20past%20ten%20years,been%20paralysed%20by%20wild%20poliovirus (accessed November 27, 2023)

33. https://www.cancer.org/cancer/types/breast-cancer/about/how-common-is-breast-cancer.html (accessed November 28, 2023)

34. American Cancer Society. *Breast Cancer Facts & Figures 2022-2024*. Atlanta: American Cancer Society, Inc. 2022.

35. https://www.cancer.org/cancer/types/breast-cancer/screening-tests-and-early-detection/mammograms/limitations-of-mammograms.html (accessed July 5, 2024).

36. https://www.cancer.org/cancer/types/breast-cancer/risk-and-prevention/breast-cancer-risk-factors-you-cannot-change.html (accessed December 2, 2023).

37. See Chapter 1 of Kampourakis, K. (2021). *Understanding Genes*. Cambridge: Cambridge University Press.

38. https://www.cdc.gov/cancer/breast/young_women/bringyourbrave/hereditary_breast_cancer/brca_gene_mutations.htm#:~:;text=All%20women%20have%20BRCA1%20and, having%20the%20same%20gene%20mutation (accessed December 18, 2023)

39. See Yates, K. (2021). *The Maths of Life and Death: Why Maths Is (Almost) Everything*. London: Quercus, Chapter 2.

40. Gigerenzer, G. (2014). *Risk Savvy: How to Make Good Decisions.* New York: Viking, p.14.

41. Gigerenzer, G., Hertwig, R., Van Den Broek, E., Fasolo, B., & Katsikopoulos, K. V. (2005). "A 30% chance of rain tomorrow": How does the public understand probabilistic weather forecasts? *Risk Analysis: An International Journal, 25*(3), 623–629.

42. Levitin, D. (2016). *A Field Guide to Lies and Statistics: A Neuroscientist on How to Make Sense of a Complex World.* London: Penguin, pp.97–103.

43. Goldenberg, M. J. (2021). *Vaccine Hesitancy: Public Trust, Expertise, and the War on Science.* Pittsburgh: University of Pittsburgh Press, pp.151–152.

44. https://www.pbs.org/wgbh/frontline/article/paul-offit-a-choice-not-to-get-a-vaccine-is-not-a-risk-free-choice/ (accessed November 24, 2023).

45. Glass, R. I., & Parashar, U. D. (2014). Rotavirus vaccines—balancing intussusception risks and health benefits. *The New England Journal of Medicine, 370*(6), 568–570.

46. Vrieze, J. (2019). First malaria vaccine rolled out in Africa—Despite limited efficacy and nagging safety concerns. *Science.* https://www.science.org/content/article/first-malaria-vaccine-rolled-out-africa-despite-limited-efficacy-and-nagging-safety; accessed February 13, 2024.

Chapter 6

1. Lea Thompson, Transcript DPT: VACCINE ROULETTE, Broadcast, April 19, 1982, 8:00-9:00 pm, WRC TV, Washington, DC, National Broadcasting Company, INC. pp.1, 3.

2. Lea Thompson, Transcript DPT: VACCINE ROULETTE, Broadcast, April 19, 1982, 8:00-9:00 pm, WRC TV, Washington, DC, National Broadcasting Company, INC, p.33.

3. Lea Thompson, Transcript DPT: VACCINE ROULETTE, Broadcast, April 19, 1982, 8:00-9:00 pm, WRC TV, Washington, DC, National Broadcasting Company, INC, p.34.

4. DTaP and Tdap are vaccines introduced in the 1990s and 2000s, respectively, that provide protection against tetanus, diphtheria, and whooping cough. Upper case letters in these abbreviations indicate that the vaccine contains full-strength doses of that part of the vaccine. The lower case "d" and "p" in Td and Tdap mean that these vaccines use smaller doses of diphtheria and whooping cough.

5. The "a" in DTaP indicates that the pertussis component of the vaccine is acellular, meaning it contains purified components of the pertussis bacterium rather than the whole cell. This acellular component reduces the risk of side effects compared to the older whole-cell pertussis vaccine.

6. https://www.cdc.gov/vaccines/hcp/vis/vis-statements/dtap.pdf (accessed December 15, 2023). The "a" in DTaP stands for "acellular," meaning that the whooping cough component contains only parts of the bacteria instead of the whole bacteria.

7. https://www.cdc.gov/pertussis/surv-reporting.html (accessed December 15, 2023)

8. Kulenkampff, M., Schwartzman, J. S., & Wilson, J. (1974). Neurological complications of pertussis inoculation. *Archives of Disease in Childhood, 49*(1), 46–49.

9. Stewart, G. T. (1977). Vaccination against whooping cough: Efficacy versus risks. *Lancet, 1*, 234–237, p.234.

10. Baker, J. P. (2003). The pertussis vaccine controversy in Great Britain, 1974–1986. *Vaccine, 21*(25–26), 4003–4010; Miller, D. L., Ross, E. M., Alderslade, R., et al. (1981). Pertussis

immunization and serious acute neurological illness in children. *British Medical Journal,* *282,* 1595–1599.

11. Miller, D. L., Ross, E. M., Alderslade, R., et al. (1981). Pertussis immunization and serious acute neurological illness in children. *British Medical Journal 282,* 1595–1599.

12. Pollack, T. M., & Morris, J. (1983). A 7-year survey of disorders attributed to vaccination in North West Thames Region. *The Lancet, 1,* 753–757.

13. Ray, P., Hayward, J., Michelson, D., Lewis, E., Schwalbe, J., Black, S., ... & Davis, R. (2006). Encephalopathy after whole-cell pertussis or measles vaccination: Lack of evidence for a causal association in a retrospective case–control study. *The Pediatric Infectious Disease Journal, 25*(9), 768–773.

14. Berkovic, S. F., Harkin, L., McMahon, J. M., Pelekanos, J. T., Zuberi, S. M., Wirrell, E. C., Gill, D. S., Iona, X., Mulley, J. C., & Scheffer, I. E. (2006). De-novo mutations of the sodium channel gene *SCN1A* in alleged vaccine encephalopathy: A retrospective study. *The Lancet Neurology, 5*(6), 488–492.

15. https://www.nobelprize.org/prizes/medicine/2008/summary/ (accessed December 28, 2023)

16. Montagnier, L. (2002). A history of HIV discovery. *Science, 298*(5599), 1727–1728.

17. Gallo, R. C. (2002). The early years of HIV/AIDS. *Science, 298*(5599), 1728–1730.

18. Additional evidence that HIV was the cause of AIDS came from the establishment of its ability to infect T-cells with the CD4 antigen, its consistent isolation from AIDS patients of different origins, and the isolation of similar viruses that cause AIDS in nonhuman primates such as macaques. This was further supported by the later isolation of HIV type 2 in West African patients with AIDS, as well as by the clinical efficacy of drugs that specifically inhibit HIV enzymes and by the fact that mutations in one of the coreceptors for HIV (CCR5) make some people highly resistant to HIV infection and AIDS. Gallo, R. C., & Montagnier, L. (2003). The discovery of HIV as the cause of AIDS. *New England Journal of Medicine, 349*(24), 2283–2285.

19. Duesberg, P. (1988). HIV is not the cause of AIDS. *Science, 241*(4865), 514–514.

20. Blattner, W., Gallo, R. C., & Temin, H. M. (1988). Blattner and colleagues respond to Duesberg. *Science, 241*(4865), 514–517.

21. Cohen, J. (1994). The Duesberg Phenomenon: A Berkeley virologist and his supporters continue to argue that HIV is not the cause of AIDS. A 3-month investigation by Science evaluates their claims. *Science, 266*(5191), 1642–1644, p.1642.

22. Duesberg, P. H. (1989). Human immunodeficiency virus and acquired immunodeficiency syndrome: correlation but not causation. *Proceedings of the National Academy of Sciences,* *86*(3), 755–764.

23. Duesberg, P. H. (1991). AIDS epidemiology: inconsistencies with human immunodeficiency virus and with infectious disease. *Proceedings of the National Academy of Sciences,* *88*(4), 1575–1579.

24. For an overview of the disagreements about the cause of AIDS, see Kalichman, S. C. (2009). *Denying AIDS: Conspiracy Theories, Pseudoscience, and Human Tragedy.* New York: Copernicus Books; Nattrass, N. (2012). *The AIDS Conspiracy: Science Fights Back.* New York: Columbia University Press.

25. https://www.nasonline.org/member-directory/members/40362.html (accessed December 28, 2023)

26. Booth, W. (1989). AIDS paper raises red flag at PNAS. *Science, 243*(4892), 733–733.

27. Duesberg, P. (1989). Duesberg's PNAS paper. *Science, 243*(4895), 1125–1125.

28. Boffey, P. M., "A Solitary Dissenter Disputes Cause of AIDS," *New York Times*, January 12, 1988, C-3. https://archive.nytimes.com/www.nytimes.com/library/national/science/aids/011288sci-aids.html.

29. Nattrass, N. (2012). *The AIDS Conspiracy: Science Fights Back.* New York: Columbia University Press, pp.77–80, 88–90.

30. https://cfs.hivci.org (accessed December 28, 2023)

31. Daston, L. (2023). *Rivals: How Scientists Learned to Cooperate.* New York: Columbia Global Reports, pp.12–13, 124–128.

32. http://duesberg.com/about/pdintroduction.html (accessed December 29, 2023)

33. https://www.latimes.com/archives/la-xpm-1991-05-21-vw-2451-story.html (accessed December 29, 2023)

34. Oreskes, N. (2021). *Why Trust Science?* Princeton: Princeton University Press, pp.59–60, 64.

35. Oreskes, N. (2004). The scientific consensus on climate change. *Science, 306*(5702), 1686–1686.

36. Cook, J., Nuccitelli, D., Green, S. A., Richardson, M., Winkler, B., Painting, R., ... & Skuce, A. (2013). Quantifying the consensus on anthropogenic global warming in the scientific literature. *Environmental Research Letters, 8*(2), 024024.http://iopscience.iop.org/article/10.1088/1748-9326/8/2/024024 (accessed January 23, 2024).

37. Powell, J. (2017). Scientists reach 100% consensus on anthropogenic global warming. *Bulletin of Science, Technology & Society, 37*(4), 183–184.

38. Anderson, E. (2011). Democracy, public policy, and lay assessments of scientific testimony. *Episteme, 8*(2), 144–164.

39. Rollin, B. E. (2006). *Science and Ethics.* Cambridge: Cambridge University Press.

Chapter 7

1. Wakefield, A. J. (1998). Author's reply. Autism, inflammatory bowel disease, and MMR vaccine. *The Lancet, 351*(9106), 908.

2. The story presented in this chapter is based on Deer, B. (2020). *The Doctor Who Fooled the World: Science, Deception, and the War on Vaccines.* London: Scribe, and Offit, P. (2010/2008). *Autism's False Prophets: Bad Science, Risky Medicine, and a Search for a Cure.* New York: Columbia University Press.

3. Wakefield, A. J., Murch, S. H., Anthony, A., Linnell, J., Casson, D. M., Malik, M., Berelowitz, M., Dhillon, A. P., Thomson, M. A., Harvey, P., Valentine, A., Davies, S. E., & Walker-Smith, J. A. (1998). Ileal-lymphoid-nodular hyperplasia, non-specific colitis, and pervasive developmental disorder in children. *The Lancet, 351*(9103), 637–641, p.637.

4. Wakefield, A. J., Murch, S. H., Anthony, A., Linnell, J., Casson, D. M., Malik, M., Berelowitz, M., Dhillon, A. P., Thomson, M. A., Harvey, P., Valentine, A., Davies, S. E., & Walker-Smith, J. A. (1998). Ileal-lymphoid-nodular hyperplasia, non-specific colitis, and pervasive developmental disorder in children. *The Lancet, 351*(9103), 637–641, p.641.

5. Chen, R. T., & DeStephano, F. (1998). Vaccine adverse events; causal or coincidental? *The Lancet, 351*(9103), 611–612.

6. Afzal, M. A., Minor, P. D., Begley, J., Bentley, M. L., & Armitage, E. (1998). Absence of measles-virus genome in inflammatory bowel disease. *The Lancet, 351*(9103), 646–647.

7. Chadwick, N., Bruce, I. J., Schepelmann, S., Pounder, R. E., & Wakefield, A. J. (1998). Measles virus RNA is not detected in inflammatory bowel disease using hybrid capture and reverse transcription followed by the polymerase chain reaction. *Journal of Medical Virology, 55*(4), 305–311.

8. http://news.bbc.co.uk/2/hi/uknews/politics/1719666.stm (accessed December 4, 2023)

9. Goldacre, B. (2008). *Bad Science.* London: Fourth Estate, p.301.

10. Boyce, T. (2007). *Health, Risk and News: The MMR Vaccine and the Media.* New York: Peter Lang, Chapter 9, quotation on p.193.

11. Wakefield, A. J., Harvey, P., & Linnell, J. (2004). MMR—responding to retraction. *The Lancet, 363*(9417), 1327–1328.

12. https://www.theguardian.com/society/2010/feb/02/lancet-retracts-mmr-paper (accessed December 4, 2023).

13. https://www.theguardian.com/society/2010/feb/02/lancet-retracts-mmr-paper (accessed December 4, 2023).

14. Horton, R. (2004). *MMR: Science and Fiction, Exploring the Vaccine Crisis.* London: Granta Publications, p.23.

15. Horton, R. (2003). *Second Opinion: Doctors, Diseases and Decisions in Modern Medicine.* London: Granta Books, p.208.

16. Chen, R. T., & DeStephano, F. (1998). Vaccine adverse events; causal or coincidental? *The Lancet, 351*(9103), 611–612, p.612.

17. Horton, R. (2004). *MMR: Science and Fiction, Exploring the Vaccine Crisis.* London: Granta Publications, p.23.

18. Horton, R. (2003). *Second Opinion: Doctors, Diseases and Decisions in Modern Medicine.* London: Granta Books, pp.209–210, 213–214.

19. https://retractionwatch.com/the-retraction-watch-leaderboard/top-10-most-highly-cited-retracted-papers/ (accessed December 4, 2023).

20. https://retractionwatch.com/2010/08/03/why-write-a-blog-about-retractions/ (accessed December 4, 2023).

21. https://www.nature.com/articles/d41586-023-03974-8 (accessed December 18, 2023).

22. Fanelli, D. (2009). How many scientists fabricate and falsify research? A systematic review and meta-analysis of survey data. *PLoS One, 4*(5), e5738.

23. Silberzahn, R., & Uhlmann, E. L. (2015). Crowdsourced research: Many hands make tight work. *Nature, 526*(7572), 189–191.

24. Gary Smith (2023). *Distrust: Big Data, Data-Torturing, and the Assault on Science.* Oxford: Oxford University Press, pp.20, 146, 111.

25. Kerr, N. L. (1998). HARKing: Hypothesizing after the results are known. *Personality and Social Psychology Review, 2*(3), 196–217.

26. Head, M. L., Holman, L., Lanfear, R., Kahn, A. T., & Jennions, M. D. (2015). The extent and consequences of p-hacking in science. *PLoS Biology, 13*(3), e1002106.

27. Bartolucci, A. A., Tendera, M., & Howard, G. (2011). Meta-analysis of multiple primary prevention trials of cardiovascular events using aspirin. *American Journal of Cardiology, 107*(12), 1796–1801.

28. Ellis, P. D. (2010). *The Essential Guide to Effect Sizes: Statistical Power, Meta-Analysis, and the Interpretation of Research Results.* Cambridge: Cambridge University Press, Chapter 1.

29. Rosnow, R. L., & Rosenthal, R. (2003). Effect sizes for experimenting psychologists. *Canadian Journal of Experimental Psychology, 57*(3), 221–237.

30. Ioannidis, J. P. A. (2005). Why most published research findings are false. *PLoS Medicine*, 2(8), e124.

31. National Academies of Sciences, Engineering, and Medicine. (2019). *Reproducibility and Replicability in Science*. Washington, DC: The National Academies Press.

32. Ioannidis, J. P. A. (2016). Why most clinical research is not useful. *PLoS Medicine*, 13(6), e1002049.

33. Ioannidis, J. P. A. (2018). All science should inform policy and regulation. *PLoS Medicine*, 15(5), e1002576, p.1.

34. Schoenfeld, J. D., & Ioannidis, J. P. (2013). Is everything we eat associated with cancer? A systematic cookbook review. *American Journal of Clinical Nutrition*, 97(1), 127–134.

Chapter 8

1. https://www.govinfo.gov/content/pkg/CHRG-107hhrg84605/html/CHRG-107hhrg84605.htm (accessed August 23, 2022).

2. https://www.govinfo.gov/content/pkg/CHRG-107hhrg84605/html/CHRG-107hhrg84605.htm (accessed August 23, 2022).

3. https://www.cdc.gov/vaccinesafety/concerns/thimerosal/index.html (accessed August 23, 2022).

4. Hviid, A., Stellfeld, M., Wohlfahrt, J., & Melbye, M. (2003). Association between thimerosal-containing vaccine and autism. *Journal of the American Medical Association*, 290, 1763–1766.

5. Price, C. S., Thompson, W. W., Goodson, B., Weintraub, E. S., Croen, L. A., Hinrichsen, V. L., ... & DeStefano, F. (2010). Prenatal and infant exposure to thimerosal from vaccines and immunoglobulins and risk of autism. *Pediatrics*, 126(4), 656–664.

6. https://blog.oup.com/2013/11/correlation-is-not-causation/ (accessed August 25, 2022).

7. Taylor, L. E., Swerdfeger, A. L., & Eslick, G. D. (2014). Vaccines are not associated with autism: An evidence-based meta-analysis of case-control and cohort studies. *Vaccine*, 32, 3623–3629.

8. Gadad, B. S., Li, W., Yazdani, U., Grady, S., Johnson, T., Hammond, J., ... & German, D. C. (2015). Administration of thimerosal-containing vaccines to infant rhesus macaques does not result in autism-like behavior or neuropathology. *Proceedings of the National Academy of Sciences*, 112(40), 12498–12503.

9. Coyne, J. (2015). *Faith vs. Fact: Why Science and Religion are Incompatible*. New York: Viking, pp.202–204 (emphasis in the original).

10. https://www.salon.com/2011/01/16/dangerousimmunity/ (accessed August 23, 2022)

11. https://childrenshealthdefense.org/events-news/unique-challenge-to-the-media-and-american-people/ (accessed August 23, 2022).

12. I am indebted to Kevin McCain for the example and the formulation.

13. https://www.youtube.com/watch?v=7VGs2PCHc (accessed August 24, 2022).

14. https://www.cdc.gov/vaccinesafety/concerns/thimerosal/index.html (accessed August 23, 2022).

15. https://www.cdc.gov/ncbddd/autism/data/index.html#data (accessed August 23, 2022).

16. https://www.salon.com/2015/09/17/donaldtrumpsdangerousignoranceonvaccinesandautismantivaxxerswonsciencelostatlastnightsgopdebate/ (accessed August 23, 2022) As Trump's twitter account is suspended it is no longer possible to access the actual tweets he made.

17. Autism and Developmental Disabilities Monitoring Network Surveillance Year 2008 Principal Investigators. (2012). Prevalence of autism spectrum disorders—autism and developmental disabilities monitoring network, 14 sites, United States, 2008. *Morbidity and Mortality Weekly Report: Surveillance Summaries*, *61*(3), 1–19.

18. Sandler, B. P. (1951). *Diet Prevents Polio*. The Lee Foundation for Nutritional Research.

19. Sandler, B. P. (1941). The production of neuronal injury and necrosis with the virus of poliomyelitis in rabbits during insulin hypoglycemia. *The American Journal of Pathology*, *17*(1), 69.

20. Sandler, B. P. (1951). *Diet Prevents Polio*. Milwaukee, WI: The Lee Foundation for Nutritional Research. https://www.seleneriverpress.com/images/pdfs/DIET%20PR EVENTS%20POLIO.PDF (accessed January 22, 2024).

21. https://www.newspapers.com/image/789035732/?clipping_id=121538793&fcfToken= eyJhbGciOiJIUzI1NiIsInR5cCI6IkpXVCJ9.eyJmcmVlLXZpZXctaWQiOjc4OTAzNTcz MiwiaWF0IjoxNzA2MDA0MjIxLCJleHAiOjE3MDYwOTA2MjF9.JQ77MAH9kQR FNXZ5Bd1_puabqPe-s_F0j7NWnZbHgRY (accessed January 22, 2024).

22. Messerli, F. H. (2012). Chocolate consumption, cognitive function, and Nobel laureates. *New England Journal of Medicine*, *367*, 1562–1564, p.1562.

23. Messerli, F. H. (2012). Chocolate consumption, cognitive function, and Nobel laureates. *New England Journal of Medicine*, *367*, 1562–1564, p.1564.

24. Maurage, P., Heeren, A., & Pesenti, M. (2013). Does chocolate consumption really boost Nobel award chances? The peril of over-interpreting correlations in health studies. *The Journal of Nutrition*, *143*(6), 931–933.

25. Flaherty, E., Sturm, T., & Farries, E. (2022). The conspiracy of Covid-19 and 5G: Spatial analysis fallacies in the age of data democratization. *Social Science & Medicine*, *293*, 114546.

26. https://www.bbc.com/news/newsbeat-52395771 (accessed July 16, 2024).

27. https://hbr.org/2015/06/beware-spurious-correlations (accessed April 3, 2023).

Chapter 9

1. https://content.time.com/time/health/article/0,8599,1721109,00.html (accessed December 29, 2023).

2. https://content.time.com/time/health/article/0,8599,1721109,00.html (accessed December 29, 2023).

3. https://www.cbsnews.com/news/vaccine-case-an-exception-or-a-precedent/ (accessed December 29, 2023).

4. https://www.uscfc.uscourts.gov/sites/default/files/opinions/CAMPBELL-SMITH. POLING041008.pdf (accessed December 29, 2023).

5. Offit, P. A. (2008). Vaccines and autism revisited—the Hannah Poling case. *New England Journal of Medicine*, *358*(20), 2089–2091.

6. https://www.hrsa.gov/sites/default/files/hrsa/vicp/vicp-fact-sheet.pdf (accessed December 29, 2023).

7. https://www.hrsa.gov/sites/default/files/hrsa/vicp/vicp-stats-12-01-23.pdf (accessed December 29, 2023).

8. Kirkland, A. (2016). *Vaccine Court: The Law and Politics of Injury*. New York: New York University Press, p.6.

9. https://www.hrsa.gov/vaccine-compensation/about (accessed December 29, 2023).

10. Kirkland, A. (2016). *Vaccine Court: The Law and Politics of Injury.* New York: New York University Press, p.2.
11. Kirkland, A. (2016). *Vaccine Court: The Law and Politics of Injury.* New York: New York University Press, p.7.
12. Poling, J. S., Frye, R. E., Shoffner, J., & Zimmerman, A. W. (2006). Developmental regression and mitochondrial dysfunction in a child with autism. *Journal of Child Neurology, 21*(2), 170–172.
13. https://www.uscfc.uscourts.gov/sites/default/files/opinions/CAMPBELL-SMITH. POLING041008.pdf, p.3.
14. https://www.uscfc.uscourts.gov/sites/default/files/opinions/CAMPBELL-SMITH. POLING041008.pdf, pp.4–5.
15. https://www.uscfc.uscourts.gov/sites/default/files/opinions/CAMPBELL-SMITH. POLING012811.pdf, p.2.
16. https://www.uscfc.uscourts.gov/sites/default/files/opinions/CAMPBELLSMITH. %20DOE77082710.pdf, p.2.
17. http://www.uscfc.uscourts.gov/omnibus-autism-proceeding (accessed January 4, 2024)
18. http://www.uscfc.uscourts.gov/sites/default/files/autism/EXITING_GUIDANCE_TO_PRO_SES.pdf
19. Gorski, D. (2008). The Hannah Poling case and the rebranding of autism by antivaccinationists as a mitochondrial disorder. https://sciencebasedmedicine.org/on-the-rebranding-of-autism-as-a-mitochondrial-disorder-by-antivaccinationists/ (accessed January 4, 2024). The full text of the concession was also published and is still available here: https://www.ageofautism.com/2008/02/full-text-autis.html
20. https://www.healthchoicevt.com/2018/10/12/10-12-2018-who-was-child-doe-77/ (accessed January 4, 2024)
21. Harris, G. (2008). Deal in an autism case fuels debate on vaccine. *The New York Times,* March 8, 2008. https://www.nytimes.com/2008/03/08/us/08vaccine.html (accessed January 4, 2024).
22. Kirby, D. Hannah Poling Really Did Change Everything, AGE OF AUTISM, (June 18, 2008, 4:27 PM), https://www.ageofautism.com/2008/06/hannah-poling-r.html (accessed January 4, 2024).
23. https://www.washingtonpost.com/archive/lifestyle/magazine/2003/08/03/reaction/ c16b42a3-ac4d-4341-b385-6fe06981d4ea/ (accessed January 4, 2024).
24. https://www.newsweek.com/robert-f-kennedy-jr-vaccines-covid-dr-fauci-i-read-science-1572688 (accessed January 4, 2024).
25. https://www.hrsa.gov/vaccine-compensation/faq (accessed January 4, 2024).
26. Kirkland, A. (2016). *Vaccine Court: The Law and Politics of Injury.* New York: New York University Press, Chapter 5.
27. Offit, P. A. (2008). Vaccines and autism revisited—the Hannah Poling case. *New England Journal of Medicine, 358*(20), 2089–2091.
28. Reiss, D. R., & Heap, R. (2018). Using and misusing legal decisions: Why anti-vaccine claims about NVICP cases are wrong. *The Minnesota Journal of Law, Science & Technology, 20,* 191–260.

29. Dickson, E. (2019). A guide to 17 anti-vaccination celebrities. *Rolling Stone*. https://www.rollingstone.com/culture/culture-features/celebrities-anti-vaxxers-jessica-biel-847779/ (accessed January 26, 2024).

30. Willon, F., & Mason, M. (2015). California Gov. Jerry Brown signs new vaccination law, one of nation's toughest. https://www.latimes.com/local/political/la-me-ln-governor-signs-tough-new-vaccination-law-20150630-story.html (accessed January 26, 2024).

31. Bueno, A. (2015). Jim Carrey slams California vaccine law. CBS News. https://www.cbsnews.com/news/jim-carrey-slams-california-vaccine-law/ (accessed January 26, 2024).

32. https://www.cdc.gov/vaccinesafety/concerns/thimerosal/index.html (accessed January 26, 2024).

33. Centers for Disease Control and Prevention (CDC). (1999). Thimerosal in vaccines: A joint statement of the American Academy of Pediatrics and the Public Health Service. *MMWR. Morbidity and mortality weekly report, 48*(26), 563–565.

34. UNESCO/World Commission on the Ethics of Scientific Knowledge and Technology (COMEST) (2005). *The Precautionary Principle*. Paris: UNESCO.

35. Offit, P. (2010/2008). *Autism's False Prophets: Bad Science, Risky Medicine, and a Search for a Cure*. New York: Columbia University Press, pp.60–70.

36. Halsey, N. A. (1999). Limiting infant exposure to thimerosal in vaccines and other sources of mercury. *JAMA, 282*(18), 1763–1766, p.1764.

37. Halsey, N. A. (1999). Limiting infant exposure to thimerosal in vaccines and other sources of mercury. *JAMA, 282*(18), 1763–1766, p.1766.

38. https://www.nytimes.com/2002/11/10/magazine/the-not-so-crackpot-autism-theory.html (accessed February 21, 2024).

39. Halsey, N. A., & Goldman, L. (2001). Balancing risks and benefits: *Primum non nocere* is too simplistic. *Pediatrics, 108*, 466–467.

40. Offit, P. (2010/2008). *Autism's False Prophets: Bad Science, Risky Medicine, and a Search for a Cure*. New York: Columbia University Press, p.73.

41. Institute of Medicine (2004). *Immunization Safety Review: Vaccines and Autism*. Washington, DC: The National Academies Press, p.1

42. Galton, F. (1865). Hereditary talent and character. *Macmillan's Magazine, 12*, 157–166, 318–327.

43. Galton, F. (1883). *Inquiries into Human Faculty and Its Development*. New York: Macmillan & Co.

44. See Kampourakis, K. (2024). *How We Get Mendel Wrong, and Why It Matters: Challenging the Narrative of Mendelian Genetics*. New York: CRC Press.

45. Goddard, H. H. (1912). *The Kallikak Family*. New York: Macmillan Company.

46. Barker, D. (1989). The biology of stupidity: Genetics, eugenics and mental deficiency in the inter-war years. *The British Journal for the History of Science, 22*(3), 347–375.

47. Stern, A. M. (2011). From legislation to lived experience: Eugenic sterilization in California and Indiana, 1907–79. In Lombardo, P. A. (Ed.), *A Century of Eugenics in America: From the Indiana Experiment to the Human Genome Era*. Bloomington & Indianapolis: Indiana University Press, 95–116.

48. Laughlin, H. H. (1922). *Eugenical Sterilization in the United States.* Chicago: Psychopathic Laboratory of the Municipal Court of Chicago, p.338.

49. Laughlin, H. H. (1922). *Eugenical Sterilization in the United States.* Chicago: Psychopathic Laboratory of the Municipal Court of Chicago, p.339.

50. Morgan, T. H. (1924). Human inheritance. *The American Naturalist, 58*(658), 385–409.

51. Morgan, T. H. (1925). *Evolution and Genetics,* 2nd edition. London: Oxford University Press, p.207.

52. Myerson, A. (1925). *The Inheritance of Mental Diseases.* Baltimore: Williams & Wilkes, p.77.

53. Myerson, A. (1925). *The Inheritance of Mental Diseases.* Baltimore: Williams & Wilkes, p.78.

54. Myerson, A. (1925). *The Inheritance of Mental Diseases.* Baltimore: Williams & Wilkes, p.79.

55. Deposition of Harry H. Laughlin, Assistant Director of Eugenics Record Office of Carnegie Institute; 11/12/1925; Buck v. Bell (Case File #31681) (accessed July 8, 2024)

56. https://supreme.justia.com/cases/federal/us/274/200/ (accessed July 8, 2024)

57. For the Buck story see Lombardo, P. A. (2022). *Three Generations, no Imbeciles: Eugenics, the Supreme Court, and Buck v Bell.* Baltimore, MD: Johns Hopkins University Press; Catte, E. (2021). *Pure America: Eugenics and the Making of Modern Virginia.* Arcadia Publishing.

58. Rothwell, J. T., Cojocaru, A., Srinivasan, R., & Kim, Y. S. (2024). Global evidence on the economic effects of disease suppression during COVID-19. *Humanities and Social Sciences Communications, 11*(1), 1–14.

59. Riou, J., Hauser, A., Fesser, A., Althaus, C. L., Egger, M., & Konstantinoudis, G. (2023). Direct and indirect effects of the COVID-19 pandemic on mortality in Switzerland. *Nature Communications, 14*(1), 90.

60. Steele, M. K., Couture, A., Reed, C., Iuliano, D., Whitaker, M., Fast, H., ... & Silk, B. J. (2022). Estimated number of COVID-19 infections, hospitalizations, and deaths prevented among vaccinated persons in the US, December 2020 to September 2021. *JAMA Network Open, 5*(7), e2220385–e2220385.

61. Sah, P., Vilches, T. N., Pandey, A., Schneider, E. C., Moghadas, S. M., & Galvani, A. P. (2022). Estimating the impact of vaccination on reducing COVID-19 burden in the United States: December 2020 to March 2022. *Journal of Global Health, 12,* 03062.

Chapter 10

1. https://twitter.com/DarlaShine/status/1095679791154581506 (accessed December 2, 2023).

2. https://twitter.com/DarlaShine/status/1095680122194182144 (accessed December 2, 2023).

3. https://twitter.com/DarlaShine/status/1095765065821810689 (accessed December 2, 2023).

4. https://clark.wa.gov/public-health/measles-investigation (accessed December 2, 2023).

5. https://edition.cnn.com/2019/03/07/health/measles-josh-nerius/index.html (accessed December 2, 2023).

6. Pew Research Center, May 2023, "Americans' Largely Positive Views of Childhood Vaccines Hold Steady".

7. https://www.cdc.gov/measles/elimination.html (accessed December 2, 2023).

8. Biggerstaff, M., Cauchemez, S., Reed, C., Gambhir, M., & Finelli, L. (2014). Estimates of the reproduction number for seasonal, pandemic, and zoonotic influenza: a systematic review of the literature. *BMC Infectious Diseases*, *14*(1), 1–20.

9. Del Rio, C., Malani, P. N., & Omer, S. B. (2021). Confronting the delta variant of SARS-CoV-2, summer 2021. *Journal of the American Medical Association*, *326*(11), 1001–1002.

10. Guerra, F. M., Bolotin, S., Lim, G., Heffernan, J., Deeks, S. L., Li, Y., & Crowcroft, N. S. (2017). The basic reproduction number (R_0) of measles: A systematic review. *The Lancet Infectious Diseases*, *17*(12), e420–e428.

11. https://www.who.int/news-room/fact-sheets/detail/measles (accessed May 8, 2024)

12. https://edition.cnn.com/2019/04/15/opinions/measles-cases-rise-global-crisis-fore-ghebreyesus/index.html ((accessed December 19, 2023).

13. https://www.who.int/europe/news/item/22-02-2024-rapid-measles-outbreak-response-critical-to-protect-millions-of-vulnerable-children (accessed February 24, 2024).

14. https://atlas.ecdc.europa.eu/public/index.aspx (accessed January 24, 2024).

15. https://edition.cnn.com/2019/03/07/health/measles-josh-nerius/index.html (accessed December 2, 2023).

16. Hutchinson, S. J. (Ed.). (1901–1903). *Atlas of Illustrations of Clinical Medicine, Surgery and Pathology (a Continuation of the Atlas of Pathology)*. London: John Bale, Sons & Danielson Limited, Plate O.

17. See Pennock, R. T. (2000). *The Tower of Babel: The Evidence Against the New Creationism*. Cambridge, MA: MIT Press.

18. Daeschler, E. B., Shubin, N. H., & Jenkins Jr., F. A. (2006). A Devonian tetrapod-like fish and the evolution of the tetrapod body plan. *Nature, 440*, 757–763; Shubin, N. (2008). *Your Inner Fish: The Amazing Discovery of Our 375-Million-Year-Old Ancestor*. London: Penguin Books.

19. Shubin, N. H., Daeschler, E. B., & Jenkins, F. A., Jr. (2006). The pectoral fin of Tiktaalik roseae and the origin of the tetrapod limb. *Nature, 440*, 764–771; Ahlberg, P. E. (2018). Follow the footprints and mind the gaps: A new look at the origin of tetrapods. *Earth and Environmental Science Transactions of the Royal Society of Edinburgh, 109*(1–2), 115–137.

20. Ellenberg, J. (2015). *How Not to be Wrong: The Power of Mathematical Thinking*. Penguin, pp.3–8.

Conclusions

1. 2024 Edelman Trust Barometer Global Report. https://www.edelman.com/trust/2024/trust-barometer.

2. Rudolph, J. L. (2023). *Why We Teach Science (and Why We Should)*. Oxford: Oxford University Press.

3. Osborne, J., & Pimentel, D. (2022). Science, misinformation, and the role of education. *Science, 378*(6617), 246–248.

Index

204